BOMB
GIRLS

JACKY HYAMS

BOMB GIRLS

BRITAIN'S SECRET ARMY: THE MUNITIONS WOMEN OF WORLD WAR II

JOHN BLAKE

Published by John Blake Publishing Ltd,
3 Bramber Court, 2 Bramber Road,
London W14 9PB, England

www.johnblakepublishing.co.uk

www.facebook.com/Johnblakepub facebook
twitter.com/johnblakepub twitter

First published in hardback in 2013

ISBN: 978-1-78219-442-2

British Library Cataloguing-in-Publication Data:

A catalogue record for this book is available from the British Library.

Design by www.envydesign.co.uk

Printed in Great Britain by CPI Group (UK) Ltd

1 3 5 7 9 10 8 6 4 2

Papers used by John Blake Publishing are natural, recyclable
products made from wood grown in sustainable forests.
The manufacturing processes conform to the environmental
regulations of the country of origin.

Every attempt has been made to contact the relevant copyright-holders,
but some were unobtainable. We would be grateful if the
appropriate people could contact us.

CONTENTS

Acknowledgements   vii

Foreword   ix

Introduction   xvii

The Poems   1

1. Building the Secret War Machine   5

2. Working in a Bomb Factory   23

3. Betty's Story: The Yellow Ladies   49

4. Margaret's Story: The Process Worker in   71
   'A' Section

5. Ivy's Story: The Girl with the Lathe   89

6. Laura's Story: An Angel and a Rose   107

7. Margaret's Story: The Teasing Girl   123

8. Maisie's Story: Maisie From Essex   137
   With the Facepack

9. Alice's Story: Fancypants   153

10. Dorothy's Story: The Government Inspector   167

11. Iris's Story: The Girl on the Bicycle   181

12: The Factories   195

ACKNOWLEDGEMENTS

My most sincere thanks to everyone involved in the All Party Parliamentary Group for Greater Recognition for Munitions Workers, namely Rob Flello, Russell Brown, Huw Irranca-Davies and Phil Wilson, who were so generous with their time in the initial stages of my research.

A special thank you must also go to Clare Tillyer at John Blake Publishing, whose infectious enthusiasm for the idea kickstarted the book into life. Also to Lloyd Brown, Chief of Staff in Rob Flello's office and Kellie Haste, Huw-Irranca Davies's caseworker at Ogmore, for all their help and efforts on behalf of the women.

Mike Clubb, author of *The Welsh Arsenal* (essential reading for anyone interested in the history of munitions work in the Bridgend factory) gave me invaluable assistance in guiding me towards the surviving munitions workers, as well as important insights into the record keeping of accidents and fatalities at Bridgend.

Thanks too must go to Vera Barber in Bishop Auckland for keeping the flame alive for the memory of the 'Aycliffe Angels' over many years and maintaining a mini-archive of material from those times.

The enthusiastic contribution of the families of the Bomb Girls I interviewed also made it so much easier to organise and conduct my interviews. David and Eunice Jagger, Matt Green, Don Curtis and Jo Street were extremely helpful every step of the way, particularly when it came to the details of their family history. Thanks too to Juliet Young at Stanway Primary, Colchester, for assistance in tracking down Margaret and Don Curtis.

And finally, and most importantly, my thanks to the memorable, brave women whose stories are told in these pages: Betty, the two Margarets, Ivy, Dorothy, Maisie, Alice, Iris and Laura. All gratitude for the time we spent together looking back at your lives – and your amazing and invaluable contribution to the history of this country.

FOREWORD

This book tells a story of heroes, ordinary women who responded to the call to serve their nation in extraordinary and dangerous times. But this is no work of fiction. Their often gruelling experiences should already be well-known and a source of national pride. Yet for nearly 70 years, these women's stories have largely gone untold.

Over the last few years we have seen recognition for a number of groups of veterans of the Second World War, most notably on the Home Front. The Bevin Boys and Women's Timber Corps in 2007, the Women's Land Army in 2009 and the participants in the Arctic Convoys in early 2013 have all received long-deserved recognition of their service.

Yet, perhaps surprisingly, there remains a group of veterans, numbering well over a million people, who remain without any form of recognition. The munitions workers of the Second World War, the majority of whom were women, were carrying out highly skilled, dirty, and extremely dangerous work, with the threat of accident or enemy action a constant threat. It goes without saying that the consequences of an explosion in a munitions factory could be horrific.

An enormous explosion at the RAF Fauld underground munitions storage depot in Staffordshire of 1944 killed 81 people, and the 1940 bombing of the Vickers aircraft factory at Brooklands, near Weybridge in Surrey resulted in the deaths of 86 workers, with 630 people injured. At Royal Ordnance Factory (ROF) sites across the country, workers toiled day and night amidst risks such as these to provide the munitions Britain needed to maintain the war effort.

But perhaps because the highly secretive nature of the work during the war bred in them the habit of not talking about their wartime experiences, or maybe because the attitude of so many meant that they didn't feel the need to draw attention to themselves, this ever-dwindling number of people in their late eighties and nineties now find themselves the true unsung heroes of the Second World War.

This book explains in vivid detail the trials and tribulations facing women in munitions factories, be it the ever-present risk of explosion, the effect that the chemicals they worked with would have on their bodies, or simply the day-to-day difficulties of living in a strange place and dealing with the restrictions of wartime. Theirs is an incredible story and one which has taken far too long to tell.

Even I, as a member of the campaign for recognition, was not aware that my partner's grandmother had worked in a munitions factory until recently. A member of my staff mentioned the campaign to a friend, who also told him that a relative of his had worked in munitions.

I know of other members of the campaign who have only recently discovered that their relatives were former workers, thanks to their involvement with the campaign. It is likely that taking into account the sheer numbers of workers in the munitions factories, most people in Britain will be related to, or will know someone who worked in the factories during the war years. Many are perhaps just not aware of it, or do not fully appreciate its significance.

At the outbreak of the First World War, Britain found herself woefully underprepared and in no position to provide anywhere near the number of shells required by the British Army. This 'Shell Crisis' was a major contributory factor to the fall of the Liberal Government in 1915, which in turn led to the formation of a coalition Government and the appointment of David Lloyd George, future Prime Minister, as Minister of Munitions. The Government came to realise that the only way to win the war would be to move the economy onto a war footing and prioritise the manufacture of munitions.

Women had previously worked in munitions factories but the changes made by the Munitions of War Act in 1915 not only made it a penal offence for a munitions worker to leave their employment in a munitions factory, but also led to an increase in the number of female workers from 212,000 in 1914 to nearly 1 million by war's end in 1918. It is estimated that female workers were responsible for around 80 per cent of all the weapons and ammunition used by the British Army in the First World War.

Keen to learn from such lessons, the Government ensured that the infrastructure of munitions manufacture was improved in the years before the outbreak of the Second World War.

Despite this, war still necessitated a huge increase in production and in the number of munitions factories. By 1943, there was even a munitions factory under Parliament, manufacturing fuse parts and parts for submarines. Once again, with men conscripted to serve in the Armed Forces, it was women who provided the bulk of the munitions workforce.

Without their work, the Spitfires defending the skies of southern England during the Battle of Britain would have had no bullets; the D–Day landing forces would have been lacking in tanks and artillery; and the Navy shadowing the North Atlantic convoys would have had no shells.

While the USA, as the 'great arsenal of democracy', certainly helped turn the tide of the war in its later years, munitions workers in British factories were crucial to the supply of our troops through the war. It is no exaggeration to say that Britain would not have been victorious without them.

This book, however, does not simply tell of an important period in our military history but perhaps equally in our social history. With the South and East of England at serious risk from air attack, munitions production moved to the North of England, Wales and Scotland and became a truly national effort. Many of those women, who had moved hundreds of miles away from their homes to work in the factories, chose to stay after the war, permanently transforming several areas around the country.

The attempts by these young women to retain some sense of normality in their lives in what was a far from normal situation are fascinating in themselves. It never ceases to amaze to hear a 90-year-old woman speak so matter-of-factly about experiences ranging from the horrifying to, dare I say, the scandalous. Yet after the war ended, as had been the case in the First World War, women did not necessarily expect to return to their pre-war occupations, having contributed so fundamentally to the war effort.

My own interest in this subject stems from a meeting in 2009 with a constituent whose mother required my help with a benefits issue. She was missing a hand. Most of her other hand was missing too. My constituent explained this was due to an explosion during her time at ROF Swynnerton, just a few miles from my constituency.

This lady had spent the vast majority of her adult life, and had brought up her children, with the most debilitating of injuries. Yet she had not received so much as a letter of thanks for her work. This struck me as fundamentally wrong. As a result, I joined with a number of other MPs who had campaigned locally on this issue to set up the All Party Parliamentary Group (APPG) on Recognition of Munitions Workers.

We have steadily grown in numbers as we have spread the word to other MPs and made them aware of the work carried out in or near to their own constituencies. We have had meetings with Ministers in an attempt to persuade them to support our campaign but disappointingly, citing cost issues and the difficulties in finding accurate records of those individuals who worked in the munitions factories, successive Governments have not felt able to provide any form of recognition.

The APPG feels that these issues are surmountable. As a result, our goal is simple: to find some means of recognising munitions workers, men and women, of the two World Wars. It is envisaged that this will be in the form of a permanent memorial at the National Memorial Arboretum in

Staffordshire and some form of badge or medal for individual workers, together with a letter of recognition signed by a representative of both the Government and the Royal Family.

With the help of a successful fundraising campaign, we hope to be in a position to commence with both of these in the near future. If, after reading this book, you want to help with our campaign, then do get in touch. There are many ways you can do so, ranging from fundraising for our ongoing efforts to simply putting us in touch with surviving munitions workers in your area.

Much of our work now is establishing exactly who worked in these factories and any help with that would be greatly appreciated. You can visit our website at www.munitionsworkers.wordpress.com.

I am delighted to have been able to help Jacky Hyams make contact with some of the munitions workers who feature in this book. Her passion for the stories that these ladies tell, and her commitment to do them justice, has been clear to see from our first meeting and this book is testament to that fact. I hope that the munitions workers of the Second World War will consider this book a fitting reflection of their work.

I know for Jacky, meeting these ladies has been both a joy and a privilege, as it has been for me, and I know that she has spent countless hours talking to them about their work. Their stories are in equal parts tragic, uplifting and amusing, but most of all simply fascinating. Theirs was truly a different world to what we know today, with young women thrust into dangerous work on the frontline of Britain's war effort, worried about their partners or relatives fighting thousands of miles away, in many cases living hundreds of miles from home in unfamiliar areas of the country. Yet amazingly, every single former worker I have spoken to, when asked whether they would do it again, has responded, without hesitation, that they would.

Most of all, it is hard not to feel a sense of shame when meeting these ladies that nearly 70 years on we still have not, as a nation, given our thanks to these brave individuals who risked their lives daily to ensure we could win the fight against fascism.

Words alone are insufficient to describe the debt which we owe to these workers but I want to pay tribute to them all – those who have contributed to this book, those I have met during the course of the campaign, those who survive and those who have passed away in recent years, and those who gave their lives in the service of their nation during the war. All made a huge contribution to the Allied victory in the Second World War. And all are deserving of a nation's gratitude.

This book tells their story and I hope it will bring us another step closer to the recognition that these heroines so richly deserve.

Rob Flello, MP
Secretary, APPG on Recognition of Munitions Workers
May 2013

INTRODUCTION

Courage in the Second World War – not the defiant, angry, in-your-face bravery of the captured resistance heroine, or the awesomely daring exploits of the Battle of Britain pilots, or those who gave all for Bomber Command – the people in this story represent a very different kind of courage. A quiet, steady, some would say typically British stoicism in the face of adversity.

Yet their bravery in wartime was a national secret. They didn't wear uniforms to demonstrate their role in the war effort; they weren't allowed to. The places where they worked each day were shrouded in secrecy of the highest order. As a consequence, like the millions of uniformed men and women whose combined efforts helped win the war, they too were forbidden to talk about what they did. Their day-to- day work was so crucial to the survival of the country, only their loved ones and colleagues could know about it. And even then, most of these brave people were only aware of the set or specific tasks they were required to perform each day.

While the nation itself lived through a time when all information was censored, these wartime heroes worked under conditions where strictly enforced secrecy dominated everything they did. They had an important job to do, but no questions could be asked.

So who were they? They weren't codebreakers, nor were they spies or espionage agents working undercover. They didn't form part of some elite group operating behind the scenes. Yet effectively, theirs was a very significant hidden army, mostly female, teenage girls, young war brides, ordinary mums, war widows and even, in some instances, grandmothers.

As Winston Churchill described them in a radio broadcast of 1940, their role was defined as 'soldiers with different weapons – but the same courage'. Unquestionably, without the work of this hidden female army, the story of the Second World War might have ended in a very different way.

But why, you might ask, are the people of this hidden army perceived as such brave fighters? After all, everyone in Britain during WW2 had no choice but to hold their nerve for nearly seven years, enduring much deprivation, devastation, separation and sorrow as a bomb-blasted land shakily made its way towards peacetime. The need for fortitude, certainly, existed for everyone, wherever they were.

The reality was, as part of the Home Front (the mobilisation of Britain's civilian population to support the war effort) these women calmly faced even more danger than the ordinary civilian. The nature of their jobs, the work they did around the clock for many years, put them right in the firing line, their only 'weapons' being their tireless ability to keep going – and the strength of their relationships with each other.

Theirs was a world where the slightest accident or a tiny mistake or slip-up at work had the potential to blow them all to smithereens and leave Britain's war effort in jeopardy – because these women worked in the country's munitions or explosives factories. Their job was to help build the bombs, fill the bullets, create the spare parts for the ammunition and firepower the country needed so badly – 'Bomb Girls' whose role, until recently, went unheralded in the history of Britain's fight for national survival, even though their numbers were estimated at 1 million-plus throughout the war.

Danger of the very worst kind was ever-present in a vast wartime munitions factory. Everyone working on the factory floor handling highly toxic chemicals risked their health. Each day carried the risk of sudden, accidental explosions causing disfigurement, blindness or loss of limbs – or worse.

As Phil Wilson, Labour MP for Sedgefield, told me: 'They went to work in the morning and came home at night – but sometimes they didn't come back at all.' Phil, whose grandmother filled shells at the munitions factory at Aycliffe, is one of a small cross-party group of British MPs campaigning for greater recognition for Britain's munitions women for many years.

My own involvement with the Bomb Girls came about after reading a brief newspaper story in 2012. For the very first time in history, a small group of munitions workers from across Britain, now in their eighties and nineties, had marched in London alongside all the other wartime veterans on Remembrance Sunday.

In some cases, their stories and memories of their wartime experiences in munitions factories had been told locally. There were memorials to the munitions heroes in a few places, yet there had never been any national, wider acknowledgement of the significance of their work in WW2.

A phone call to the British Legion, who put me in touch with the organisers of the Remembrance Sunday munitions march, revealed the long silence surrounding the Bomb Girls and their work was about to end. A cross-party group of MPs from around the country was now overseeing a big campaign for greater recognition for the munitions workers. With their help, I was able to contact a number of surviving Bomb Girls across the country, all happy to talk to me about their lives and experiences. After war ended, they'd all returned to normal life, married, raised families and put it all behind them. Yet when I sat down to talk to them, the memories came flooding back without any prompting: their recall of those years working on the factory floor seemed as vivid as if it had happened just a few years back.

After my initial interviews with two such workers, Betty Nettle in Bridgend and Margaret Proudlock in Dumfries, whose stories were so fascinating and well-told, it was difficult to tear myself away, I asked the campaigning MPs why it had taken so long for these women to be vocal about the work they'd done for their country. Phil Wilson told me he believed it was that British national stoicism, the quiet 'get on with it' courage of millions, which contributed to the women's long silence. Many of them were no strangers to tough times, either.

'They thought their stories were not unique because it was happening to everyone else they knew,' he says. 'Everyone faced the same dangers. And in many of the areas the munitions women came from, adversity was not unknown – pit, farm, or Armed Forces – those were the employment choices where they lived.'

Russell Brown, MP for Dumfries & Galloway and part of the cross-party campaign, knows munitions work well. Before becoming an MP, he worked for ICI at the Nobel's Explosives Factory at Powfoot, Dumfries, for 23 years. Many of those years were spent in munitions work. He believes a lack of record keeping in wartime might also be one of the reasons why formal recognition has been so slow in coming.

'At a time of war, perhaps paperwork wasn't the most important issue of the day,' he told me. 'I made initial approaches in 2008 to say we should recognise the munitions veterans but frequently all that came back from the then DTI [Department of Trade & Industry] was "we have no records of these people".

'It's obvious how dirty, heavy and dangerous the work was way back then, compared to the seventies and eighties when I worked at ICI. But I suspect there were many incidents in factories across the country where people were badly injured or killed, yet some seem to have been reported, some don't. Safety wouldn't have been the priority that it is today in many places. Robust accident records? It's questionable.'

Yet there was more to contend with than physical danger. As I interviewed more munitions women, other less obvious factors emerged. The pain of separation from loved ones was common enough in wartime. Communication was limited, mostly, to letter writing. But consider the emotional effect on young women not even out of their teens, girls like Maisie Jagger, 'called up' at 18 to work in munitions, then sent off by the authorities to a different part of the country, far away from home.

Maisie, an East Ender, told me the separation from her family and home made her physically ill. She was literally pining for home, fading away. Thankfully, the authorities recognised this and eventually allowed her to work near her home.

For Laura Hardwick, who has the distinction of being both an Aycliffe 'Angel' and a Swynnerton 'Rose', being sent to live away from her home in the Northeast to live nearly 200 miles away, in a purpose-built hostel, was an experience she struggled to endure with considerable stoicism, until the months before war ended, when depression overwhelmed her.

The final straw, she told me, was when her friend and roommate returned home to Scotland: like so many other women, only the bonds of friendship helped Laura get through the sheer, exhausting slog of it all. And it was a slog. The factories ran 24 hours a day, seven days a week.

The women's shifts were rotated round the clock. For some Bomb Girls, living several miles away from their workplace, a 12-hour day, including travel time, was common. The worst times, they all told me, were the night shifts in the blackout. Bombing raids from above created more than just a risk of lives – they meant reduced production hours. While security and safety rules were everywhere, there were still times when nothing, not even a bombing raid, was allowed to stop the production line. It was that hairy.

Rob Flello, MP for Stoke on Trent South, who helped

launch the all-party munitions campaign in 2011, reminded me of the stark reality of those times: 'People sat in the train during an air raid in total silence for what seemed like hours in the darkness. Even the railway station that they used every day didn't officially exist.'

For the women, most of whom had left school at age 14 to work in poorly paid domestic service, the money they earned in munitions was a huge incentive. But the relentless, exhausting routine, the danger of working constantly with toxic chemicals and the ongoing threat from the enemy in the skies above exacted a crushing toll. How on earth, you wonder, did they hold it together?

Huw Irranca-Davies, MP for Ogmore in Wales, joined forces with Rob Flello in 2011 to launch the recognition campaign for munitions workers. He confirmed what Laura Hardwick and others told me: it was the relationships the women forged with those working with them that made the difference. The strong sense of community in remote areas like Bridgend, the huge munitions factory where Betty Nettle and thousands of other women worked, played a big part.

'In many ways, it was the making of that part of mid Glamorgan, after the war,' he said. 'What was special about that community is that it was particularly Welsh in the way they came together, masses of people coming together – and having to make the best of what they were given.'

'It was just a job, they didn't think of the danger,' said Vera Barber, from Bishop Auckland. Vera has been very closely involved locally in keeping the memory of the Aycliffe munitions women alive for many years. Strictly speaking, Vera wasn't a Bomb Girl. She worked at the Aycliffe munitions complex in the administration section, as a teenage clerk.

For the office workers it was different, a 9 till 5 job, no night shifts. Yet Vera still remembers friends at work who were killed in accidents, and the many women in factories who faced the worst personal tragedies, but kept going regardless. 'They might have lost a loved one the day before, but they still went to their shift the next day. You never heard the word "stressed".'

Vera too believes it was the closely knit ties of the local communities that sustained everyone, got them through. 'People just got on with it, especially northerners. We're tough lot up here.'

These women's stories are, if you like, brief snapshots of wartime, glimpses of what now seems an almost unimaginable scenario: family life torn apart, bombs falling from the sky, everything in people's lives closely regulated by the authorities – including their war work.

Yet the other fascinating aspect of the Bomb Girls stories is their collective social history, the background to their lives. Many described childhoods growing up happily, in conditions that we would today regard as deprivation. And each woman's story underlines how very different family life was then.

Parental authority, for instance, was not to be questioned under any circumstances. Nearly all the women told me they had originally hankered for a role in the Armed Forces. Yet their parents were firmly against it because the perception, at the time, was that the Forces were not suitable for a respectable young woman. Munitions work was seen (ironically) as a safer, softer option that paid well, too.

It would be difficult to recount these women's stories out of context. The first two chapters of this book describe the background to the creation of the big munitions factories, the hazards therein and an overview of the way munitions factories ran through the wartime years.

For readers wanting to know more about the factory sites themselves, the final chapter gives brief factual details of the sites where the women in the book found themselves working. In the spring of 2013, after a gap of 70 years, Iris Aplin and her friend Mary Taylor were filmed by the BBC revisiting the Swynnerton site where they'd worked as teenagers – a poignant reminder of their personal history, and the history of so many others.

Nine women from a remarkable generation tell us their stories here. Talking to them at length – a real privilege for any writer – has only served to confirm what is part of our national conscience. Our debt to them, and the millions like them, is enormous. It can only be repaid by acknowledging their worth, time and again, and honouring their times – and all the sacrifices they made. That they did it so selflessly and quietly, without any fanfare, makes it even more remarkable.

Jacky Hyams
London
May 2013

THE POEMS

During my meeting with Vera Barber, she showed me some paperwork and other material she had kept from her years working in the offices at the Aycliffe munitions factory. I was intrigued to find, scrawled in pencil on the back of a few yellowing delivery notes from November 1944, a series of poignantly worded poems.

Vera, aged 20 at the time, told me she believed these had been written by a book-keeper around her own age, a girl called Marian Taylor. The girls had worked together for some time and Vera recalled her colleague writing the poems to help her when she learned that her young fiancé, posted overseas, had been killed.

In many ways, the existence of these poems serves to underline the camaraderie between workers that became a hallmark of the munitions factories – and the small, unobtrusive ways in which people supported each other through the bad times.

Here they are.

TO A HERO

Far away in a distant land 'neath the blazing stars
Away from all the ones he loved
A corner of England lies.
Lost to all the world around,
A hero, unhailed, unsung;
His soul we know by God is found,
And his crown of Victory won
We often wish to have seen his face,
When he was safely laid
Into his last resting place
And military honours paid.

THE MEMORY

Why do you seem to keep smiling at me?
Though your dear face I never more will see
I picture you smiling through the rosy twilight glow,
And stop to throw a kiss before you turn and go.
Although you went away so very long ago
I still recall you there, and really you must know
You'll always find me writing dear and I will never go
You always knew the answer dear, because I love you so.'

The following three short poems are untitled:

As I wing my way through the clouds
into the blue above
I take one last look at the city, the city that I love,
And whisper from my heart, I will come
back again some day
Until the last smouldering ruin has gradually
faded away.

As the notes of the Last Post thrill,
I imagine I see them salute;
And their hearts with emotion fill,
And their voices all remain mute
His short, young life is done
His earthly task is o'er;
The pounding of the gun,
His one, last farewell.

Although the logs have fallen and light is fading fast.
And still you stand before me, until I feel at last
I bear the imprint of your smile, impressed
upon me here,
Then with a tiny sigh you slowly disappear.

CHAPTER 1

BUILDING THE SECRET WAR MACHINE

T he history of Britain's female munitions workers goes back to the First World War, when women played a vital role in munitions production. Yet, 3 September 1939, the day the Second World War was declared in Britain, marks the point at which the wartime story of the women in this book began. Though in reality, preparation for the looming inevitability of war with Germany – the world war everyone hoped might never happen – was already underway.

The big warning signs came with the appointment in Berlin of Germany's new Chancellor, Adolf Hitler, in 1933. This news was received all over Europe with much trepidation: until that point, Germany's Armed Forces were not considerable, a direct consequence of the German military defeat at the end of the First World War in 1918.

Yet by the thirties, as his power gained momentum, Hitler's plans for full-scale domination of Europe and beyond quickly became all too visible. By 1934, the German Armed Forces were rapidly increasing – and the country's war machine was expanding at an incredible rate.

In Britain, the Government was initially reluctant to accept the idea of this immensely powerful threat. Who wanted war? With the shadow of the First World War still looming large after just two decades, the argument for pacifism was heard everywhere. Wouldn't it be better to wait and see what happened? Couldn't there be negotiation for peace?

These were, of course, false, flimsy hopes. It became very clear that Britain urgently needed to set about re-arming, building up its Armed Forces – and, most importantly, building brand new factories to produce the planes, the ammunition, the bombs, the guns, the bullets, everything that would be needed in the increasingly likely event of a German onslaught.

The new Government-owned munitions factories, called Royal Ordnance Factories (ROF), would be built solely to supply Britain's Armed Forces in wartime. Armaments factories like these were not new: the first ROF facilities had originally been built to increase munitions production during the First World War.

They were sited around London, in Woolwich and Enfield and Waltham Abbey in Essex. The historic Royal Arsenal at Woolwich, on the River Thames to the Southeast of London, had been, until war with Germany became imminent, Britain's principal producer of armaments. But now it was agreed that the Woolwich site was out of date. Moreover, its central location made it highly vulnerable to aerial bombardment by Hitler's airforce, the Luftwaffe, since it was known that the conflict ahead would involve bombing raids.

As a result, the all-important decision was made: brand new big armaments factories in safer, more suitable places away from the Southeast of the country had to be built, and quickly. This decision to avoid new construction in the Southeast proved to be tragically accurate. Throughout WW2, while over 30,000 people continued to work in armament production at the Woolwich Royal Arsenal, 103 factory workers were killed and over 700 injured during a series of bombing raids from V1 flying bombs and V2 rockets.

FINDING THE FACTORY SITES

The locations for the new armaments factories had to be very carefully chosen. The criteria were strict: the new sites had to be on level ground and they had to have the right geological conditions, so that some of the buildings could be built underground. The sites would be recruiting thousands of workers from within a 25-mile radius – yet for safety reasons, the ammunition factories could not be near a large centre of population. Nor could the factories be easily visible from the air.

The new factories producing the weaponry also needed to have good road and rail connections, though their existence – and their construction – would have been shrouded in the utmost secrecy.

As soon as war was declared in the autumn of 1939, all possible existing factory resources were needed, too: large private companies like ICI Nobel and Lever Brothers (known as Unilever today) and a number of other firms had to switch their normal production to manufacturing goods for the war effort, making everything from uniforms to aircrafts, ammunition and tanks, all under the auspices of the Government. Meanwhile, normal peacetime factory production was suspended, resulting in huge shortages of everyday goods for sale around the country (in addition to the food and other forms of rationing that were introduced early in 1940).

There were four different types of munitions factory:

• Engineering factories producing the metal casings for bombs and shells or, in some instances, producing parts, rifles, guns and tanks.
• Small–arms factories producing the bullet casings. (These factories were often existing engineering factories turned over to war production.)
• Explosive factories manufacturing various explosive agents.
• Filling factories to fill the bomb and shell casings with the explosives. The risky nature of working with combustible explosive material meant that these were the most dangerous of all munitions factories. In these, located right across the country, the raw ingredients of explosives, shells, casings and detonators were brought together to make bullets, shells and mortar bombs.

The four different types of munitions factories and their locations are listed below. Staff from the Woolwich Arsenal helped design and, in some instances, oversaw the construction of the new factories.

THE FILLING FACTORIES

Woolwich, London
Hereford, Herefordshire
Chorley, Lancashire
Bridgend, Glamorgan
Glascoed Usk, Monmouth
Swynnerton, Staffordshire
Risley, Lancashire
Kirby, Liverpool
Thorpe Arch, Yorkshire
Aycliffe, County Durham
Rearsby, Leicestershire
Burghfield, Reading, Berkshire
Healey Hall, Rochdale, Lancashire
Ruddington, Nottinghamshire
Walsall, Staffordshire
Elstow, Bedfordshire
Featherstone, near Wolverhampton, Staffordshire

THE ENGINEERING FACTORIES

Woolwich, London (Woolwich had a dual function as a filling factory)
Enfield, Middlesex
Birtley, County Durham
Blackburn, Lancashire
Cardiff, Glamorgan
Cardonald, Glasgow, Scotland
Dalmuir, Dumbartonshire, Scotland
Fazakerley, Liverpool

Leeds, Yorkshire
Hooton, Cheshire
Newport, Monmouthshire
Radcliffe, Lancashire
Maltby, Rotherham
Wigan, Cheshire
Patricroft, Manchester
Ellesmere Port, Cheshire
Hayes, Middlesex
Poole, Dorset
Nottingham, Nottinghamshire
Theale, Berkshire
Hirwaun, Glamorgan

THE SMALL ARMS AMMUNITION FACTORIES

Radway Green, Cheshire
Blackpole, Worcestershire
Spennymoor, County Durham
Steeton, Yorkshire

THE EXPLOSIVES FACTORIES

Waltham Abbey, Essex
Bishopton, Renfrewshire
Ardeer, Stevenson, Ayrshire
Drungans, Dumfries
Edingham, Dalbeattie, Kircudbrightshire
Pembrey, Carmarthenshire
Wrexham, Denbighshire
Drigg, Cumberland

Bridgwater, Somerset
Ranskill, Notts

A VERY DANGEROUS JOB

While all the new munitions factories were sited in areas that made them difficult to locate from the air, had the Germans been sure enough of their targets, there would have been few, if any, survivors of an air raid, given the extremely hazardous nature of working with explosives. The factories needed to take every possible precaution to minimise every risk, with the bombing threat being the most deadly. The hundreds of different chemicals used in the manufacture of weaponry and bombs made the ongoing risk of a bombing raid a terrifying proposition.

Then there were the risks involved for the workers handling the highly explosive material. Some of the chemicals used in production, like cordite – a 'low explosive' propellant comprising nitro-cellulose and nitro-glycerine, made in the form of cords or sticks and used to send projectiles to a target – were regarded as safe to handle.

Yet some of the materials used in detonators, the small copper shells used to initiate the triggering process, could only be filled by hand with very sensitive materials like lead azide, which looked similar to castor sugar, or fulminate of mercury, a highly toxic yet harmless-looking light brown powder. These explosive materials were so sensitive, they could cause injury to hands, fingers and faces during the filling process itself.

Fulminate of mercury, for instance, is sensitive to friction, heat and shock and can decompose violently into mercury, a lethal element absorbed through the skin, the lungs and the digestive system. Mercury has long been known to cause mood swings and, in extreme cases, madness. The common British expression 'mad as a hatter' to describe a crazy person, had its origin in the 18th and 19th centuries when mercury was used to manufacture felt for hats.

Making the millions of pellets that boosted the ignition of weaponry was a highly hazardous process, too. An explosive component in the pellets, called tetryl, also a highly sensitive chemical, was a yellow powder which could affect the skin and cause a number of skin complaints. Workers' skin turned yellow, leading to some of them being nicknamed 'Canaries' or 'Yellow Ladies'. Hair would turn a yellowish hue (hair was supposed to be covered at all times, but even then, the proximity of the chemicals could still cause discolouration). And the sensitivity of tetryl made it a high-risk chemical for causing accidents and explosions.

Some people experienced breathing problems and asthma from handling the many chemicals used to make the bombs and weapons. Cases of dermatitis (skin rashes) and other skin problems were much higher in filling factories than any other industrial disease. In some instances, people's teeth fell out as a direct result of the chemicals they worked with.

These were the known hazards in munitions work. Scientists tried to come up with solutions to these problems, sometimes with creams and ointments, but the priority in wartime, irrespective of all the serious hazards, remained the same: to keep the production line running continuously.

When contact with an offending chemical reduced a worker's ability to do the job, they could be taken off the production line and moved to another, less hazardous, section of the factory. Yet the nature of the production line – and the filling factories were very much round-the-clock, seven days a week operations – involved a constant balancing act: taking one worker off the production line created an urgent need for a replacement.

Throughout the years of war, some effort was made to reduce risk for munitions workers. Regular checks were introduced for employees working with certain bomb-making chemicals such as TNT (trinitrotoluene). TNT poisoning can damage the stomach lining and cause other serious complaints such as anaemia, jaundice, heart problems, liver failure and lung cancer. These checks did lead to a reduction in cases of poisoning, yet for some workers they came too late. It is not possible to say how many workers in munitions factories died as a direct result of working with chemicals.

There are no officially published updated statistics for those who perished as a consequence of chemical poisoning, however, from January 1941 to the end of July 1945, the large filling Royal Ordnance factories maintained official records of fatalities from accidents or explosions in the factories. These records show that around 90 people lost their lives in accidents in the big filling factories. But how can such official figures possibly tell the real story?

There were many munitions workers who survived beyond the WW2 years, but subsequently died from an illness related to their exposure to the chemicals they worked with. Or there were those who survived an accident but whose lives were wrecked by terrible injuries. The consequences of war, all too often, continue for decades beyond the official date of ceasefire.

The words 'small arms propellant', the description for the

explosives used in an arms factory, sound fairly innocuous. In fact, propellant is a ferocious material in an enclosed area. It destroys everything in its wake.

SECRECY AND SAFETY

Aside from the huge risks involved in working with chemicals and explosives, there were other important priorities to consider while Britain's armaments resources were being built up.

The issue of wartime security had to be strictly enforced, which meant that the level of secrecy surrounding the building and formation of Britain's munitions factories was so high, in some instances even local people living in the surrounding area were sometimes completely unaware of the factory's existence right through wartime.

Then there were also the huge day-to-day, routine safety issues behind the factory walls to be implemented. Safety would have to be everyone's responsibility, workers and bosses alike. All safety rules existed to be observed to the very letter. If not, the consequences were lives lost in the factory – and production disrupted.

There could be no margin for error in the work itself. As mind-numbingly repetitive as some of the filling factory work on the production line could be, it still required total concentration. You needed a very steady hand. Consider the work of the woman working on the production line handling tiny detonators for the bombs. Her job involved picking up the detonator – half the size of an aspirin tablet – with a pair of tweezers. One tiny mistake, the slip of a hand, could be fatal. And if the munitions were to leave the factory incorrectly assembled, the consequences for the front line troops could be fatal too. The risks were there all the time.

Once the factories went into operation, safety rules were handed out, in booklet form, to every munitions employee. At one of the largest ROF filling factories, at Aycliffe, County Durham, the Rules of the Danger Area booklet ran to over 35 pages and covered every aspect of safety at the factory, including rules concerning accidents and extinguishing fire.

These safety rules were also posted as slogans around the factory canteen. In addition, the rules were read out to the workers at the beginning of the first shift every week. Anyone breaking those rules faced dismissal or worse. In one instance, a worker who was caught smoking in a munitions factory was handed a prison sentence. That's how tough the rules were. However, human nature being what it is, rule breaking did go on at times. Pressure to keep to production line targets led to some people 'cutting corners' in the work. But it was a very dangerous game to play.

'BE LIKE DAD, KEEP MUM'

The combination of strict safety and tight security also meant that the factory workers, employed in a vast site of many different sectors, each sector producing a different type of munition, would be turning up for work, performing their assigned task on the factory floor through an eight-hour shift, eating their meals in the factory's vast canteen and even enjoying regular live entertainment that was staged in the canteens.

Yet other than during specified breaks or at lunchtime, the

majority of workers were not permitted to move around the factory building. Popping into another section to have a chat with a friend was strictly forbidden. During the blackout at night, when the slightest sign of light could alert bombers to the factory's existence, women even had to walk to the ladies' toilet with an escort. Every minute at work had to be accounted for; even that trip to the ladies would be carefully logged by a supervisor to ensure maximum production targets could be adhered to. The workers knew nothing of what else went on in their workplace beyond their immediate place of work.

No questions could be asked. Official information, like everything else at the time, was rationed and handed out when it was deemed appropriate. 'Officialdom' (ie the Government) was in control of all information, no matter how trivial, relating to the war. And the munitions factories were, effectively, an arm of the fighting forces.

'Be like dad, keep mum; careless talk costs lives' was wartime's most popular slogan, launched by the Government in 1940. It was a simple poster campaign, tirelessly directed at the British population through the war years. And its message worked. Though of course gossip or rumour outside the factory often relayed snippets of information to the workers about what had been happening.

If a worker heard a loud explosion outside their section, few would ask their supervisor or boss what had happened. Everyone working on the shop floor knew what it signified. Some newspapers carried reports of accident fatalities after the event, giving the names of those killed, but never the location of the factory itself.

So secret were the locations of these bomb factories, even RAF flyers were warned to keep away from a specific area where a site was located. They might hazard a guess, but they weren't going to be told what was going on in the sites.

Today, we are accustomed to the idea of health and safety regulations. We complain about them frequently. Yet back during WW2, if the worst happened and a munitions worker died or was badly injured in an accident or explosion, their co-workers on the production line might sometimes see, with their own eyes, the appalling truth of the dangers they all faced. Yet there was no stress counselling or support backup available to help ease the co-workers' distress or shock. The emergency services were called and frequently, whole areas would have to be cleared for everyone's safety, but it was very much a case of 'once it's all cleared up, get back to business as usual, as quickly as possible'.

Managers or supervisors would briefly acknowledge the distress of any staff witnessing a terrible accident. But that was it. Time off for stress? Unheard of. Just one workers' nervous reaction to a serious incident was seen as a risk factor that had to be contained, no matter what. Someone who couldn't 'hold their nerve' was considered a danger to the production line – and would be likely to find themselves transferred to other work elsewhere. Or in some cases the factory bosses would simply let them go.

GETTING TO WORK

The logistics of getting the thousands of munitions workers to the factories in areas that were not centrally located and shrouded in secrecy also had to be very carefully planned.

The timing had to be right: transporting large numbers from home to factory and back at exactly the right time to start or finish a shift required military-like precision.

In order to ensure that all production ran round the clock, seven days a week, workers had to adhere to a strictly controlled rotating shift pattern which meant that those involved in the filling factories usually worked from 2pm to 10pm, 10pm to 6am, or 6am to 2pm, though there were variations on these timings across the country, depending on the work involved.

Mostly, the munitions workers would start their winter working day in total darkness, rising at 4am, rain or shine, to walk a mile or two to board a bus or train. If it was a very long journey, they would be collected from a train by another bus which would finally deliver them direct to the factory gate. This commute to work could be relatively short but in some cases it might be as long as two hours each way, in addition to a normal eight-hour shift.

Consider too that even these journeys to and from work weren't always without hazard if there were air raids in the skies above: this meant that virtually every aspect of the working life of munitions workers was fraught with danger. There were specially constructed air raid shelters within the factory sites, but at times an initial air raid siren would be ignored by those in charge in order to keep the production line running just a little bit longer.

THE DANGER BUILDING MEN

Once they were transported to the factory gate and before entering the main building, the workers clocked in and showed their passes to the security men. Then they could change into their factory clothes in buildings called 'shifting houses'.

The safety rules meant that everyone was forbidden from taking certain everyday items into a munitions site. The banned items were nicknamed 'contraband' and had to be left in a special hut at the entrance to the factory site. Many things were designated as contraband, especially metal items, because the tiniest spark anywhere in the building could cause an explosion. Needless to say, matches, cigarettes and lighters were banned, but also all types of jewellery, metal hairclips, and even sweets – eating sweets with fingers that had touched explosive chemicals was highly risky.

If a married woman did not want to remove her wedding ring – and many of the women working in the filling factories were married – then it had to be covered with tape. Male workers (nicknamed as 'danger building men' by the girls) would often carry out spot checks, looking for dangerous items.

The factory clothes the workers changed into were either overalls or white jackets and trousers, with either a white turban or a turban-style headscarf to cover up hair. Even this clothing was designed so that it did not have any metal fastenings – another risk factor in an explosives factory – and the outfit was made of cotton, to reduce any risk of a spark from build-up of static electricity. No metal was permitted in the workers' footwear, either, and the women were often required to change into regulation footwear.

Yet there was very little on offer as protective clothing, the sort of thing we would expect nowadays. In some factories, barrier creams were supplied to the women to protect their skin from any effects of the chemicals they were handling, but many decided they preferred to wear their own heavy makeup – thick 'pancake' on the face – as a form of protection.

THE SECRET ARMY

Throughout the war years, the munitions workers were making a considerable contribution to the war effort, risking their own lives and safety to do an important job. Yet as far as the rest of the population were concerned, the munitions workers could not be easily identified, other than those who were unlucky enough to find themselves working with chemicals that discoloured their skin or hair. Apart from their own families, co-workers and loved ones, they were very much a secret force.

Across the country, as the war went on, uniformed service people and voluntary workers became a familiar sight everywhere. There was status involved with certain types of uniform (the RAF, the Navy) because they instantly identified the wearer as performing a specific role.

Yet the 'hidden' nature of munitions work meant that the munitions workers were often unknown warriors. They did not have any kind of distinguishing uniform to indicate that they too were helping win the war. The official reason for this was the risk of the wearer carrying any kind of contamination from their work in the factory complex out into the wider world. So they became, in a sense, invisible to the wider public, preoccupied with the hard daily slog of food rationing, blackouts, bombing raids and all the other tough, testing rigours of wartime.

All workers at Royal Ordnance Factories were given a small

badge, bearing the words 'ROF, Front Line Duty', to wear on their coats when they were not at work. Its official purpose was to signify to shopkeepers and others that the bearers were engaged in important war work, and that they should be given preferential treatment. Yet officially the factories did not exist, so very few people not connected with the sites knew about them. As a result, only a few munitions workers were given the small privileges they deserved.

This manifested itself in tiny ways. While a free cup of tea provided by the Women's Voluntary Services (WVS) would be served to the hordes of uniformed personnel waiting to board trains at railway stations, female munitions workers would have to team up with a group of soldiers and ask them to collect a cuppa for them. It sounds trivial now, but it underlines the lack of recognition of the munitions workers' efforts – and their worth – during wartime.

As a workforce, the munitions women did not effectively start working in really large numbers until two years after war broke out, when the new factories were all set up and munitions production got underway. When war was declared, there were just seven factories producing ammunition. By 1941 there were 44 munitions sites in operation right across the country and eventually this would total 66.

Many of these factories employed very large numbers of workers of both sexes. The newly created Welsh Arsenal, the huge ROF filling site at Bridgend in Glamorgan, known as the largest munitions site in Europe, employed 32,000 workers at its peak in 1942; Swynnerton, the big ROF filling site in Staffordshire, employed 18,000 workers at one point; and Aycliffe, the vast filling site in County Durham employed 17,000 people.

Here were huge enterprises, working nonstop, employing a large 'Secret Army' who didn't stop to question it all: their long working hours, the dangerous nature of the work or the secrecy that surrounded them. They just got on with the work. And the majority of these workers were women, an estimated million-plus of them, working throughout the war, filling the bombs, making the bullets, assembling the jeeps, repairing planes, testing tanks. Their job was to make sure that Britain's fighting forces – their own brothers, sweethearts or husbands – had the ammunition to fight the enemy and win the war.

To describe their work as hazardous is something of an understatement. The Bomb Girls endured much: exhaustion, fear, sacrifice, separation from loved ones, personal or family tragedy – not to mention the enormous risks to their own lives and physical welfare as they worked. Yet these women, all ages, married or single, from different backgrounds, were a crucial link in the long chain that made up Britain's wartime endeavour.

The men were sent off to fight, fire the bullets, drive the jeeps, fly the planes and drop the bombs. But it was the Bomb Girls who helped make the final victory possible. They too were amongst the country's true heroes of wartime.

CHAPTER 2

WORKING IN A BOMB FACTORY

When war was declared in 1939, the idea of women getting directly involved with war work was not, at first, a serious consideration. Back then, a married woman's role was to remain at home, looking after the family. It was generally assumed, at first, that life would continue very much as before.

Single women worked mostly in 'women's jobs' in an office or a shop. Or they worked in service, or as nurses or teachers. Once a woman got married, her working life usually ended. Moreover, at the outset of war, it was firmly believed that men would deeply resent any idea of their wives or daughters being 'called up' or conscripted (having to register for compulsory military service) into any kind of war work.

Yet even before war actually broke out, women were ready to get involved. Many had already signed up for part time voluntary war work. At the time war broke out, the WVS (Women's Voluntary Service, now known as the Women's Royal Voluntary Service) already had 165,000 members and their earliest task was to help in the evacuation of over one-and-a-half million mothers and children from the cities at risk of bombing raids to the comparative safety of the countryside.

In due course, the WVS went on to provide much-needed support for the thousands of people in Britain whose homes were bombed or destroyed. As uniformed volunteers, their work was to organise rest centres, food, clothing for bomb victims, and a great deal more. By 1941, their ranks had swelled to over a million.

Other voluntary women's work was to be found in the NAAFI (the Navy, Army and Airforce Institutes) running clubs, bars, shops and other facilities for servicemen and their families. Throughout the war, 96,000 NAAFI volunteers ran over 10,000 outlets, including 900 mobile shops.

Yet war has a nasty habit of rapidly destroying the established social order. In October 1939, every man between the ages of 18 and 41 not working in a 'reserved' or essential occupation (an important job necessary to the country's survival) was required to register for service in the Army, Navy or Airforce. Britain's 'call-up' had started.

THE WOMEN ARE CALLED UP

Within a year, it became clear to the Government that without women getting involved in paid war work there would be a severe shortage of essential labour. With so many men away fighting, women were urgently needed to take their place. Everything pointed towards non-voluntary conscription.

So while large numbers of women continued in their voluntary work through the war years – somehow managing to fit the unpaid work in with running the home and looking after their family – other women were gradually drafted-in to paid war work.

Historically, women had worked in paid work in wartime. During the First World War, 1,600,000 women had been paid to work in a wide range of 'men's jobs' – as bus drivers, post office clerks as well as more traditional women's jobs like teaching and nursing. The munitions factories of the time employed 950,000 women, all of them risking their lives and health to help the war effort. But these 'modern women' – or 'munitionettes', as they were called at the time – came forward and voluntarily signed up for the work; they were not conscripted.

So the idea of women being called up proved to be controversial, even shocking to some: servicemen and politicians were vocal in their dismay at the idea of 'men's work' being taken over by women. There was outcry too at the very idea of women joining the Armed Forces (the Army, in particular, was regarded in those days as too sleazy or rough for 'decent' women). Neither did the idea of women stepping in to work in reserved occupations appeal to many people.

Yet the imperatives of war were rapidly overtaking such doubts. There was a huge labour shortage; the munitions industry desperately needed an extra 1.5 million workers. The authorities had already been forced to conscript men out of essential factory work to send them into the Forces, now the ammunition was urgently needed for the troops, and machines could only do so much. Someone had to stand on the production line, fill the bombs, make the bullets, produce the spare parts and help make the war machines to ensure the troops had what they needed to fight a world war.

The official Government line was that they preferred to encourage voluntary conscription, but in the end the desperate need for workers won. Through various government initiatives and the Essential Work Order (which meant that the State was in charge of all recruitment) the first women's call-up for war work came in March 1941, a decision which would eventually affect every able-bodied, available woman in the country between the ages of 18 and 50.

In a speech in March 1941, the then Minister for Labour, Ernest Bevin, made an urgent appeal to women to come forward for war work, mainly in shell-filling factories. He said he did not want women to wait for registration to take effect; he wanted an immediate response, especially from those who might not have been in employment before. But he warned: 'I have to tell the women that I cannot offer them a delightful life. They will have to suffer some inconveniences. But I want them to come forward in the spirit of determination to help us through.'

Women did, however, have a choice. They could either sign up for the women's sections of the Armed Forces – the ATS (Auxiliary Territorial Service), the WAAF (Women's Auxiliary Air Force), or the WRNS (Women's Royal Naval Service commonly known as 'the Wrens') – or they could register to work in a reserved occupation, in a factory or in farming.

Single women aged 20 and 21 were the first to be called up. By the end of 1941, women up to the age of 30 were also required to register for either the Armed Forces or factory/farming work. In due course, the call-up was extended to women up to age 50. These successive groups of women had to register for the call-up at their local labour or employment exchange. Across the country, groups of women stood patiently on the street, waiting for their turn to register. Announcements giving details of the registration date for each specific age group were printed in the newspapers. Everyone knew they had to do it.

So great was the need for women workers that eventually, when trained nurses were in very short supply, any woman up to age 60 had to register to be called up. Former textile workers up to age 55 were also conscripted. (Wives of men serving in the Armed Forces, however, were regarded as exempt from any war work that might take them away from home.)

Single women, however, were categorised as 'mobile'. Under the Government's Emergency Powers (Defence) Act, any able-bodied person could be sent anywhere as part of the war effort: the Government could order them to leave home and go wherever their labour was needed. Their travelling expenses would be paid by the Government and they would be given help with their accommodation needs and training if required.

If training could not be given on site, mobile workers were relocated to an appropriate training centre for a set period of time before they could start their job and their accommodation was arranged for them. At the beginning of the war, there were special eight or 12 weekday and evening training courses set up by local government in London for those willing to pay a fee (£1 2s 2d) to learn how to be munitions workers. Women came from as far as Scotland to attend the courses. (The cover of this book depicts a group of women attending one of these courses in Lambeth, South London.)

What this all meant was that huge numbers of women were being relocated to another part of the country, a daunting prospect for the thousands who had never left the confines of their local area. Though for a few, of course, the chance to get away to a new environment and be free of the confines of family restriction was welcomed – as was the pay-packet.

WOMEN'S WAR WORK BEGINS

Between 1939 and 1945, more than 500,000 women served in Britain's Armed Forces. They were not permitted to serve in battle, but they took over important supporting Forces roles as drivers, radar operators, medical orderlies, working on anti-aircraft posts, as military police and wireless operators and in many other different types of important jobs.

Britain then was still a class-bound society. The women who opted to join the Armed Forces tended, mostly, to be from upper- and middle-class backgrounds – the WRNS was regarded as the most exclusive force, the ATS the least appealing, given the Army's bad reputation. So in most cases, it was the women from ordinary working-class backgrounds who opted to sign up for factory or farm work.

The idea of working on the land held much appeal for some more free-spirited women who liked the prospect of an outdoor life. The Women's Land Army (WLA) was originally formed during WW1 and it re-formed in 1939. More commonly known as the 'Land Girls', and initially a voluntary service, the WLA totalled 80,000 women workers in WW2. In fact, this proved to be a very tough option, with back-breaking 70-hour working weeks and very poor pay from the farmers employing them: £1 85s a week, which was increased to £2 85s a week in 1944.

The consequence of all this was that large numbers of women were undertaking jobs they might never have been considered for previously, even at the time war was declared. Within a short period of time, women were working everywhere: in munitions, aircraft factories, repairing planes, as mechanics or engineers, helping build ships, driving fire engines and ambulances – or turning up on night duty as air raid wardens.

At war's peak, 7 million married women were engaged in full or part-time war work in Britain, and 90 per cent of single women were employed in munitions factories. In so many ways, war work, despite the blackout perils and shortages experienced in day-to-day life, carried in its wake a newfound economic independence for all women.

Certainly, they had no other option once female conscription had been introduced. Yet in areas with high levels of unemployment, the chance to work, earn, and play a role in helping win the war was far from unwelcome. (In the event, some women voluntarily opted to work in munitions factories even before the call-up.) Yet whatever their circumstances, all involved tackled their assigned jobs with determination and gusto. This was a time when people really were 'all in it together'. Every person's contribution mattered.

RESERVED OCCUPATIONS

At the outbreak of WW2, these were the skilled or important jobs necessary for Britain's survival:
* Doctors
* Teachers and university lecturers
* Farmers, agricultural workers and students of agriculture
* Scientists
* Police
* Merchant sailors
* Railway workers
* Dock workers
* Utility (water, gas, electricity) workers
* Certain sectors of the Civil Service and local authorities
* Certain types of engineering and factory work
* Miners (this was not a reserved occupation at the beginning of the Second World War, but in 1943, conscripts were sent to work in the coalmines alongside experienced workers, giving birth to the Bevin Boys, named after Ernest Bevin, then Minister of Labour).

UPHEAVAL AND SEPARATION

In September 1939, Britain was a very different place to the world we know today. Your position in the social pecking order defined your position in life, usually for good. And Britain was not, by any stretch of the imagination, what we'd now consider an affluent society. Unemployment and widespread poverty across the country from the late Twenties to the mid-Thirties had already blighted millions of lives. Six million homes, for instance, lacked an inside toilet, an equal number even lacked a hot water supply. There had been serious attempts to improve the lives of the poorest: huge slum clearance programmes had been underway in the thirties, but these ceased abruptly as wartime drew closer.

The announcement of war meant separation, upheaval and disturbance for virtually everyone in the country, rich or poor. Normal, everyday life was disrupted in so many ways. Strict rules were put in place concerning food and other types of rationing; small children were evacuated to the country to live with total strangers; husbands or sweethearts were sent off to war for unknown periods of time in far-flung countries, or captured by the enemy and incarcerated in prisoner-of-war camps.

Air raid sirens disturbed night-time sleep, heralding bombing raids that sometimes destroyed homes and lives. Hastily constructed Anderson shelters (steel air raid shelters erected by families in their back garden: cramped and cold, they did save lives) sprang up all over the country. And thousands of people, workers and members of the Armed Forces, were on the move, travelling across the country to wherever their job or role sent them.

Because the all-important munitions factories were dotted across the country, many away from the big city centres, conscripted munitions workers who had to leave their family home to work there were sometimes sent to live in workers' hostels or estates specially constructed close to the factories. Or they tried to find 'digs' – rented rooms in other people's homes. For many young women, conscription into munitions work some distance away meant travelling to unfamiliar surroundings – amongst a sea of unfamiliar faces.

This was more emotionally difficult at the time than we could imagine today. Mobility for ordinary working people was quite different then: people used trains and buses for local travel or at holiday times, but in more remote rural areas, cycling or walking often tended to be the only way to get around each day.

Many women, married or single, had never ventured far beyond their home, family and locality. Car ownership for the masses was still to come; flying wasn't yet an everyday experience. So the sheer discomfort or strangeness of being moved to an unknown place, far from home, added to the other difficulties of working in wartime. Not only did many girls miss the familiarity and warmth of their family environment, the long hours they had to work (a 50-hour working week was not unknown for those working in Royal Ordnance factories) plus the pressure to keep up production quotas, working in round-the-clock conditions, meant that the Bomb Girls, like most of the population, would be an exhausted workforce by the time the war was over.

In areas with heavy industry, like the Midlands, a small number of women had already been working in factory jobs before the war began. Yet many of the younger Bomb Girls who were called up had previously only ever worked in service, perhaps as serving maids or kitchen staff for a middle- or upper-class employer, a common mode of employment then for young women, especially in rural areas.

On the plus side, the need for large numbers of women to work in munitions factories brought valuable employment to remote, deprived areas of Britain where jobs of any kind had traditionally been hard to find for many years. Moreover, the pay in munitions factories was good. The average for all women's pay in the early wartime years was around £1 12s 6d. Yet a week working in a big Royal Ordnance factory would bring in between £2 and £4 a week. With overtime and bonuses, a woman's weekly pay packet could be boosted to as much as £8. In some parts of the country it wasn't unusual for a Bomb Girl's pay packet to exceed the sum of money her father had been bringing home, causing family friction sometimes. Yet male pride, for now, had to be cast to one side. What mattered was winning the war.

WHEN HOME BECAME A HOSTEL

Munitions factories were often located in fairly isolated parts of the country, hence the need to construct special, purpose-built hostel accommodation for their workers. There were not enough rural homes in these areas to provide sufficient accommodation for the women being sent there. And even if the homes had been available, young women, away from home often for the first time, free of all ties and earning what was usually more money than they had ever seen before, could be regarded as an added responsibility for families in private houses who might consider putting them up. Officially, any munitions worker wishing to live in a private home near her place of work was free to do so — but such accommodation wasn't always available.

The workers' hostel accommodation was built close to the factories and designed to be totally self-contained. Some experienced male munitions workers and their families, relocated from one large facility to another, would be housed in specially constructed housing estates or camps close to the factory, consisting of semi-detached bungalows with two or three bedrooms.

Many of the big hostel complexes were long, low, camouflaged brick buildings with a manager in charge. A typical hostel camp could house up to 1,000 munitions workers and the complex would have had everything the workers could possibly need: a hairdressing salon, living quarters for visiting husbands or sweethearts, (though most married women preferred to request leave to go home if their husband came home), a hospital, a chapel and an air-raid shelter, fully kitted out for any emergency.

The weekly charge for this accommodation was £1 2s 6d, which included two meals a day. These meals were served to the workers, cafe style, in a large communal dining room in the main administrative building, which also housed a kitchen, a post office and a cinema. Living quarters for the women were in 'huts', built in units to accommodate between 50 and 90 women in one building. Each hut usually had its own communal living room with chairs.

The women's bedrooms were located down either side of a long corridor. At one end of the corridor were baths, showers, toilets and a small laundry. The bedrooms themselves, with flooring of plain linoleum, were just wide enough to house two women in two single beds. A bureau or chest of drawers was placed at the foot of each bed. There was also a light with a separate control switch for each bed, so that if one girl wanted to read and the other did not, they did not disturb each other. Directly opposite each bed was a window, usually with a small shelf underneath, with a washstand with hot and cold running water. A locker and a chair made up the rest of the bedroom furniture. Some girls would place the lockers side by side to create a small partition area in the middle of the room. The furniture was modern, pale wood, with short curtains for each window. In some hostel camps, even central heating was provided.

Because the munitions factory operated 24 hours a day, the hostel accommodation was organised so that everyone living in the same unit worked on the same shift, to ensure that everyone would go to sleep at the same time and the place would not be noisy with people coming and going at different hours. Each girl had responsibility for her own laundry and washing. There was even ground made available outside for gardening, should anyone wish to use it.

This was all carefully planned – but it was still communal living. And not all the hostel accommodation was built to house women workers in small bedrooms for two. Other hostel camps offered dormitory style accommodation, where a number of workers slept in the same area. As a consequence, people were thrown together in ways they might not have experienced before, sharing with others from all over the country, total strangers whose background – and day-to-day habits – might be quite different to their own.

It's likely that some girls, accustomed to comfortable, middle-class homes, would have found this kind of communal living spartan and unwelcoming. But for others – given that many homes in Britain did not even have bathrooms, showers or indoor toilets, and that many young women living in cramped accommodation at home had only ever slept in a shared bed with a sibling – the hostel accommodation was comfortable, even luxurious.

Just as at work, there were rules in the hostels: women had to be in at night at a specific time and they needed a special permit if they wanted to be out later, although some hostel managers or wardens did make the attempt to ease the severe homesickness and emotional upheaval that affected some of the women. They would do their best to ensure that women from the same area or newly made friends could be housed together at first until they got used to this strange new world away from home.

The idea of building hostels and accommodation specifically for munitions workers to be close to their workplace was a good one in theory, but in practice the scheme proved unpopular. Many women preferred a long commute each day to a lengthy separation from home and loved ones (workers' transport costs were Government subsidised by an assisted travel scheme).

As a result, not all of the hostel accommodation was used. In some areas, like Bridgend, there were many empty places. Eventually, these hostels wound up being used by other groups of war workers, such as the Land Girls, or families whose homes had been completely bombed out.

Today, the exact location of many of these hostels remains unclear, partly because WW2 records of munitions factories and hostel complexes were far from comprehensive, due to the secrecy of the entire munitions operation. However, thanks mostly to diligent local research or the historical archives of the corporate enterprises that switched over to munitions facilities during wartime, some records of the hostel complexes of WW2 do survive.

Munitions factories around the Coventry area in the Midlands, for instance, had a total of 16 purpose-built hostels, erected away from residential areas but close enough to the factories for a short daily commute. Workers living in these hostels were employed making aircraft parts, ammunition and more everyday items like braces (to hold up servicemen's trousers). After the war, the Coventry hostels accommodated all manner of workers from across the country and abroad, during the immediate post-war period when Britain was being rebuilt.

THE THINGUMMYBOB

As we've established, there was ultra-tight security in the munitions factories, and the work had many dangers, but it would be wrong to give the impression that the Bomb Girls' working life was unrelentingly exhausting and downbeat. As the war progressed, the Government recognised that in order to be fully productive and for morale to be maintained, the munitions workforce would need ongoing support and acknowledgement of their role.

So workers' safety issues were tightened up, and in the larger munitions complexes provision was made for married women's needs, with nurseries and 'shopping time' organised. There was also a need for a certain amount of respite away from the demands of the production line. So within the biggest munitions complexes – which were more like large towns with rail networks and their own road infrastructure – a wide range of leisure facilities for the workers was set up.

There were social clubs, darts teams, operatic societies, rugby and football clubs, drama groups, cinema screenings, all provided to encourage an active range of social activities away from the shop floor. All factories with more than 250 workers carrying out war work had to provide a canteen offering a fresh, hot meal of the meat-and-two-veg variety for about 10 pence (the equivalent of £1.50 today) and the workers had to be given a reasonable amount of time away from the factory floor to enjoy their meal or tea break.

Music played a very big part in the Bomb Girls' lives – as it did in wartime for the rest of the country whose main sources of entertainment were either the radio (or wireless, as it was known) or the cinema. (BBC Television started broadcasting in the thirties, but was disbanded in wartime and it wasn't until the fifties that television became an increasingly popular feature of everyday life.)

Dancing, too, became more popular than ever. Across the country, every town or village had a church or school hall where men and women could gather to dance the 'old fashioned' dances – the waltz, tango, foxtrot and the quickstep. In the big city dancehalls the youngsters would try out the newer dance crazes, especially the energetic and athletic jitterbug, imported by the hordes of lively American troops stationed in Britain in the mid-1940s.

People needed an escape and dancing fulfilled a dual function – as an accepted way of men and women pairing off, certainly, but also as a huge distraction from the sheer hard slog of living and working in wartime. Everyone needed something to look forward to. In the bigger factories, music would be played constantly in the shops from loudspeakers, and the most popular wartime radio programme of all, the BBC's *Music While You Work*, providing continuous live popular music, was broadcast twice daily, Monday to Friday. Launched in June 1940, the programme was specially aimed at factory workers and those in the Forces – though it was so successful it outlasted wartime and continued to be broadcast until 1967.

Another important BBC Radio programme that started out as a morale booster for factory workers was *Workers' Playtime*. This was broadcast at lunchtime, three days a week, live from the big canteen area of a different factory 'somewhere in Britain'. Singers, musicians and comic performers would perform their acts, watched by an audience of applauding munitions workers. Many of the performers were professionals but occasionally the broadcasts used local amateur talent.

For the broadcaster, it meant transporting crew, equipment, pianos, producers, musicians and variety artists up and down the country three times a week for performances – yet many of the popular comedians and singers of the time were involved. These included future stars such as Peter Sellers, Tony Hancock, Frankie Howard and Bob Monkhouse, as well as big-band singers of the time like *Coronation Street*'s Betty Driver, who died in 2011. *Playtime* was one of the first-ever touring variety shows on the BBC. It ran for 23 years.

Dance music by orchestras also dominated the airwaves and local dancehalls: bandleaders such as Harry Roy, Jack Payne, Geraldo, Joe Loss, Victor Silvester, Billy Cotton, Henry Hall and Mantovani were wartime household names, and the popular songs of the time – romantic and sentimental tunes that echoed the innermost feelings of millions – would frequently be heard over the radio, allowing listeners to spontaneously join in. Or small groups of munitions workers would start singing the songs together in the canteen as they relaxed between shifts, or on their daily journey to and from the factory.

Songs such as 'The White Cliffs of Dover', 'We'll Meet Again' and 'There's a Boy Coming Home on Leave' were a powerful expression of many Bomb Girls' feelings about their lives, with sweethearts or husbands in the Forces far away – and no end in sight to the conflict. Everyone knew all the words. With music and dance, they could momentarily forget about the war, lose themselves in romantic dreams and hope for a better time ahead.

Even today, so evocative are the sounds and words of those wartime songs it is very easy to understand how such a relatively innocent thing as a group of women singing their hearts out really did help to bolster flagging spirits – and deal with the difficulties they all faced.

Halfway through the war, the Bomb Girls even had a popular song written about them: 'The Thingummy Bob (that's going to win the war)' was a song about the factory worker making the parts or components for the wartime weapons. Recordings of the song by big-name entertainers like singer Gracie Fields and comedian Arthur Askey made millions smile – and reminded everyone that without such women's work, the war might never be won.

The song started:

> I'm the girl that makes the thing
> that drills the hole that holds the ring
> that drives the rod that turns the knob
> that works the thingummy bob.

In other words, my job may be boring and repetitive – but I'm an important part of the war effort.

OTHER ENTERTAINMENTS

The dance floor was the acceptable place for socialising between the sexes. (Pubs and bars were then very much a male domain: unaccompanied women weren't, as a rule, likely to go into them.)

The wealthy, or the officer classes in the Forces, had nightclubs and smart hotels or specially organised dances in the Officers' Mess for their off-duty socialising, while the Bomb Girls frequently had dances organised for them in their factory canteens or at the hostels. Women were always encouraged to invite 'a friend in khaki'.

For single girls, when their shifts permitted, an evening off at a local dance hall brought a welcome chance to dress up, get on the dance floor and meet new faces in uniform. In the remoter rural areas, spending hours getting to the dance, even if it meant cycling 10 or 12 miles or more to join in the fun, wasn't seen as an obstacle: the break from the relentless factory routine was what mattered most of all.

Even the tea break at the hostel or factory canteen was a time for the women to gather together in small groups, sit around drinking tea and chatting. Camaraderie or comradeship between friends and workers never ran so high as it did for these women through the war years. It also did much to help alleviate the loneliness of life for married women while husbands were far away.

Apart from *Workers' Playtime*, other live entertainment from ENSA (Entertainments National Service Association) was also staged in the factories as a means of maintaining factory workers' morale. ENSA was set up in 1939 by theatre producer Basil Dean specifically to entertain the troops and factory workers. Its first live show took place at the Woolwich Arsenal in July 1940 and as the war went on some of the performers – many of them big entertainment names of the day such as Vera Lynn, Tommy Trinder or Anne Shelton – visited the factories and travelled all over the world to entertain the troops.

ENSA shows weren't always a success because the quality of the concerts tended to vary – not for nothing was ENSA nicknamed 'Every Night Something Awful' – but the shows still gave workers the chance to let their hair down, laugh and relax for the briefest of times. And the performers worked hard at it: at times, groups of ENSA performers would give three shows a day in canteens in the larger Royal Ordnance factories.

The local cinema too was a very popular form of escapism, with its chance to see a double bill: a major film, a 'B' movie (a low budget film, often with lesser-known actors which was, essentially, a 'two for one' value offer for the cinemagoer) and a newsreel like Pathé News.

Without television, the cinema was the only opportunity for people to see filmed news stories of the battlefront. Sometimes, the women in the audience would sit transfixed, peering at the screen, hoping for a longed-for glimpse of sons or husbands or sweethearts. Shots of the RAF shooting down German planes would be greeted with cheers and clapping.

SPENDING MONEY

As a rule, single girls living at home would hand over most, if not all, the contents of their pay packets to their mother – a time-honoured tradition in homes where the extra money was often badly needed. Because their munitions pay was better than any previous earnings, a small sum was often handed back to the wage earner as spending money. For some of the younger Bomb Girls, this represented an opportunity to go shopping.

Despite the heavy rationing restrictions – items like fabric or material for new clothes were rationed as well as food – their factory work meant many of these girls could spend money at will for the first time. (This was also the case for women who hadn't worked before but found themselves doing a part-time paid job in wartime.) The shops were far from being crammed with goods, and many items, like cosmetics, were in short supply, but having money, no matter how little, to spend as they pleased, was a novelty and gave a real uplift. Yet it was mostly the Bomb Girls in their teens and twenties who could enjoy the dances and the shopping trips during their time off, whereas women with families were more restricted.

Some women held down two or three part-time jobs through the war. This kind of juggling of part-time work had not been known for women before, yet now the women were in demand, and their employment was secure. In war work you needed official release from the job, you couldn't just hand in your notice without formal approval. Most Bomb Girls who did leave the factory job left for health reasons only, though a few of the cheekier ones did take advantage of the fact that their jobs were relatively safe and would take sick leave when it wasn't strictly necessary.

HEALTHCARE

Employing large numbers of factory workers in dangerous work also meant that on-site medical facilities were essential. Accidents took priority, of course, but the general health and wellbeing of the workers was seen as being equally important in the case of ordinary sickness like flu, gastric upsets or dental health problems. Having medical staff on site helped reduce loss of production due to sickness.

In the early war years, the factory medical facility was sometimes run by voluntary workers, but eventually qualified doctors and hospital trained staff were recruited into the bigger munitions factories, though labour shortages meant recruitment was never easy. Some factories would have one fulltime nurse on site with trained doctors on call by phone in the event of an emergency. Others, such as Bridgend, had six fulltime Medical Officers and a staff of 60 nurses, dispensers and orderlies by late 1944, as well as a fleet of ambulances.

THE OTHER KIND OF WAR...

Most munitions women got on well with their male colleagues and they would usually enjoy a joke and a laugh together. But some men, perhaps disgruntled because health problems or their age prevented them from joining the fighting forces, didn't feel very happy about this 'new order' of having women working alongside them. The men were sometimes concerned about the safety of their own jobs, especially in rural areas where unemployment had been high for years. It was an attitude along the lines of, 'I've worked hard to get here and you women think you can just come in here just like that'. It didn't help, of course. Old habits die hard.

One consequence of this resistance from a few male workers was that the special government training for certain types of jobs requiring engineering or technical knowledge was not always put into practice afterwards. A woman might undergo a four- or eight-week training course for a specific engineering role only to discover, once she started work, that she was assigned to a lower level factory job because a male supervisor or colleague didn't believe it was 'women's work'.

This was frustrating, especially as, before war started, organised training for women had mostly been restricted to domestic service work. But the engineering training itself was, nonetheless, a step forward. And Bomb Girls who were good workers were rewarded with promotion: a diligent, careful worker could be moved up to 'Blue Band' (supervisory status) sometimes being placed in charge of more than one facility (or 'shop'). Yet most of the Bomb Girls' factory work was routine, unskilled labour: the majority of women went straight from the labour exchange to the factory floor.

THE PROPAGANDA MACHINE

Poster campaigns and filmed footage of Allied victories shown in cinemas were important for the nation's wartime morale. But the need to keep war workers fully motivated, to keep munitions production at its peak, also meant passing the positive message on in other ways.

Despite all the secrecy around the day-to-day factory routine, it was clear to the authorities that the so-called secret army needed to see some form of appreciation for their efforts. The general public couldn't be told what these women were actually doing or where they were. (The newspaper captions in the 'spin' stories never gave the location of the factory.) But the authorities knew that somehow, they had to do everything possible to keep the women's motivation high.

In the bigger factories, as the war started to turn in Britain's favour, the factory's radio system was used, broadcasting 'good news' bulletins interspersed with *Workers' Playtime* and *Music While You Work*. The authorities felt, right from the beginning, that emphasis also had to be made to the workers on the significance of their role in comparison with the mind-numbing routine of the everyday toil on the production line. In today's parlance this was Government public relations, or 'spin', in the form of carefully planned events designed to keep workers' spirits up – and show the public, via the newspapers and cinema, that the workers' efforts were being supported.

Members of the Royal Family undertook visits to the munitions factories, as did the female entertainers of the times. Vera Lynn and Gracie Fields – who were adored by millions everywhere – performed in the factories for the Bomb Girls as early as 1941. These visits were usually filmed for the newsreels. One such newsworthy event came in 1942, when Britain's wartime Prime Minister, Winston Churchill, visited the Aycliffe Royal Ordnance Factory in

Staffordshire and the Aycliffe Bomb Girls. Although it was mid-May, it had been snowing in the area in the days before the visit, and the snow had turned to brown slush. This, the workers decided, was definitely not good enough for 'Winnie', whose inspirational radio broadcasts throughout the war did so much to boost the spirits of the nation. So, in order to make their site more attractive for the esteemed visitor, the Angels went out to find clean snow from the surrounding countryside, and carefully laid it on top of the slush!

In 1942, special Works Relations Officers were sent out across the country to educate the workers in the Royal Ordnance factories. These visits focused on the progress of the war itself, as a means of inspiring or encouraging workers. Events were organised, with guest speakers and educational films. On a few occasions, visits were set up to other ROF sites, so that the women from selected factory sectors could see the positive results of their efforts.

Letters from troops themselves were read out, thanking the Bomb Girls for their efforts, and on one occasion a letter was sent from the Desert Rats (the 7th Armoured Division) in North Africa, specifically thanking workers in Bridgend for 'never sending out a single dud mortar bomb'.

To a large extent, this propaganda worked well. But in the Bomb Girls' own stories of their munitions years which follow this chapter, it is obvious that the difficulties they often faced made considerable impact on their lives, both during and after wartime. They were young, innocent and inhabiting a world where everyone around them was 'doing their bit'. And even now, what comes through loud and clear in their memories of those times, was their sheer grit, their plucky resilience in the face of being conscripted to work in a job that was dangerous, exhausting and sometimes debilitating.

Their stories of wartime work underline the fact that theirs was very much a generation that didn't ask questions but just 'got on with it'. Yet only now, all those years on, can we fully recognise – and acknowledge – their worth.

CHAPTER 3

BETTY'S STORY: THE YELLOW LADIES

'ARE YOU ANYONE'S BUDGIE?'

Betty Nettle was born in 1925 and has lived in the Stormy Down/Kenfig Hill area of Bridgend, Glamorgan, her entire life. She started working at the Welsh Arsenal, ROF Bridgend, as a teenager in 1941 until war ended in 1945. Her husband of over 50 years, Ivor, died in 2005. This is her story:

Work was very scarce in this area before the war came along, other than in the mines. And we were not a mining family. My father, Leonard William Cornish Reynolds, was a ganger, working on the railway, looking after the tracks. I was the youngest girl in a family of seven children. By the time I arrived, the eldest three had already left home; they were grown up, working. My sisters Edith and Nancy went into service in a big house up in London: that was the only work option then.

Families were big in those days, so if someone had a shop,

their children worked in it – or their nieces and nephew, which meant that round here, as far as work went, it was who you knew, not what you knew.

I grew up in a respectable, double fronted house. As a child – I must have been about three – my earliest memory is of my younger brother Joe being born in the front room. (We always called him 'Joe', but his real name was Norman.) You didn't have a nurse or anything like that when a baby came, you had a local lady from along the road from us: one of those ladies that 'did'. They learned how to 'do' as they went along; they brought people into the world and they laid them out for the undertaker, so it was usually the same person that came to the house when someone was born or died.

Our local lady that 'did' was also a herbalist. We lived just outside the village, about one-and-a-half miles away, just a few scattered houses, really. You rarely saw a doctor. It was always the lady that 'did' that came round. If you were sick, she'd make you up a medicine. She'd never really tell you what she was doing; she didn't say what was in the medicine. Yet people came from a long way in our area to see her, so she must have been doing something right.

We were well shod, always plenty to eat, lots of friends around us. My mother, Elizabeth Jane, was a good manager. My dad was allowed to hunt rabbits by a local farmer, so we had chickens, ducks, geese. Everything had a home with us: cats, dogs, even people sometimes. My mother was a good cook, too. At home we had porridge or boiled eggs, and sandwiches in school – it was too far away to go home for lunch – and always a cooked meal at 5pm, when Dad came home from work.

Our home revolved around our father, which the way life was then. If my mother said 'no', it was no good going to my father. We all knew the rules and we didn't go beyond them. My mother was a churchgoer and my father's favourite saying was: 'You be careful of our name. You've got it till you marry – the boys have to carry it all their lives.' It was our good name we lived to. That was it.

In 1930, I started school. Bryndu was the name of the school. Walk a mile-and-a-half to the village, then walk another mile up the road to get to the school. We could have gone to a nearer school, but we chose to go there. My brother Jack was there, my sister Mary, and then me. As long as you behaved yourself, it was fine.

I was bright enough, I suppose. I wasn't naughty, not in school. We had a very old teacher who lived in Porthcawl. Any naughtiness and he'd be warning us: 'I'll see your father.' We didn't know if he did see him, but we didn't risk Dad's displeasure. It was my mother who was the 'flipper', who had a bit of a temper. I only saw my father lose his temper once. If you were reading something, whatever it was, my mother, if she wanted me to do something would give me a 'flip' – not exactly a smack, but you knew she meant business. I can still hear her now: 'Betty, get your 'ead out o' that tuppeny novel.'

Us kids played out everywhere – under the trees, in the woods, the fields, by water, everything a kid could wish for. I was good at sports, especially rounders [a game played with bat and ball]. And I was a good sprinter. My sisters would come home from their jobs in service once a year, just for a week. They'd get half a day off a week and that one week a year. The world they lived in seemed so far away to us then.

Today, of course, you've got that world in your living room. We had radio and the newspapers to tell us about the world. We'd listen to things like *Children's Hour* or *Dick Barton* or a variety show. But you didn't have much interest in politics because it wasn't there, in your living room, in the way it is now. I honestly don't remember my parents ever talking to us about what was happening in Germany with Hitler. So we lived with ignorance, and innocence, too. There is a real difference. You had a life to live and none of that stuff about the world concerned you in the way it does now. And I don't think you really looked to the future; the idea of the future didn't push me forward that much. As a kid, you lived in the present. Whatever else was going on was very far away.

At school, there were poor, sickly kids who would have to be given things like malt regularly [malt extract was popular in the twenties and thirties as a dietary supplement for children who might be deficient in minerals and vitamins] but my mother kept us fit, both with what we ate and certain things she would do.

For instance, every spring my mother would mix up a basinful of treacle and sulphur till you could stand your spoon in it. Each one of us would give our spoon to our mother and she would put it through the mixture. We'd line up for that spoonful; we called it medicine. In the winter, if you had a cold coming, you either got elderflower tea or a blackberry concoction with hot water and sugar to sip. Now, of course, we call those things natural remedies, buy them in a shop. But then, they were never shop-bought.

Every Monday morning, a man with a horse-drawn cart would come round; he'd sell everything you could think of – soap, cloths, tea towels, paraffin for the oil lamps. We didn't have electricity for many years. My parents were dead by the time we had electricity here, many years after the war. They were offered it, of course, but they said it was too expensive. We didn't have running water indoors for many years, either. We had a tap across the road, or we went further on to the well on the other side of the railway line. The loo was an earth closet down the bottom of the garden.

We had a 'copper' in our garden shed. [A copper was an early water heater and boiler, built into a corner of a room with space underneath for a fire to heat it.] It was made of brick with a big copper bowl. You filled it up with water, lit the fire underneath and it boiled. So you'd have a wash with the hot water every day and a bath whenever. The bath would hang on a nail outside.

Of course, this was a mining area. The coal 'lords' built the houses for the workers, though that didn't involve us, as a non-mining family. The history of Wales, of course, is all about coal – I believe the first-ever million-pound cheque was written in Cardiff and that was for coal. At one stage, much later, I remember going up to the Rhonda Valley with a friend and we'd walk up to the houses – they all looked the same, everything was the same. It was easy to walk into the wrong house!

Financially, at home, everything depended on the money the father brought into the home – and how well the mother managed the wages. There'd be gambling, drinking, womanising going on, so it really did depend on what your parents did or were up to when it came to how the family lived. I appreciate the fact that I had good parents.

In July 1939, I left school. I was 14 and my mother said to me, 'Betty, there's a job going at Kenfig Hill, minding a little one-year-old, start Monday morning.' So I literally walked out of school on the Friday and started work the following Monday. It was all arranged. My mother knew these people.

It was lovely minding the baby, a real little doll. Get him up in the morning, wash him, dress him, feed him, take him for a walk in a big boat pram. That pram was posh but the family had a drapers shop. It was actually in the front room of their house, but it was still a drapers and sold everything. I'd come away at about 6pm and the pay was half-a-crown [2s 6d] a week, a half-day off on Saturday and no work on Sunday mornings.

The day they told us we were at war, that September, my mother was crying her eyes out. She'd lost a brother in the First World War. And she had sons. My dad was too old to be called up. But of course, you couldn't see what was going to happen. We had a new aerodrome a few miles away at Stormy Down and we already knew the big arsenal was being built at Bridgend. But who knew what was ahead?

Not long before war started, my brother Jack, seven years older than me, was already helping build the arsenal. He'd worked in quarries locally before that. Of course, once word went round about the arsenal – and that working there paid more money than the quarries – off he went to work there.

Nothing much happened in those first few months of the war. Women were encouraged to volunteer and, of course, all single women were conscripted by 1941 and this was later extended to married women. But I was too young to be conscripted. My sister Mary, a year older than me, wound up working in the arsenal in pyrotechnics. My sister Nancy had married and left service, so she wound up working at a paper mill – Dickensons in Watford, near London. Just a couple of days after war had been declared she brought her little son, Leonard, down to our house – and left him with us. Nancy's husband was going into the Forces straight away. All the talk was they were going to bomb London, so Nancy left Leonard with us for safety. After that, the family always called Leonard one of the first evacuees.

I carried on working with the baby for about 18 months after war was declared, then an older person started running the drapers' shop so I found a different job, still child-minding, in the next village. I was looking after a little boy and a small baby.

By Christmas, 1941, there were uniforms everywhere. For a 16-year-old, that meant good fun at the dance. I know it sounds stupid now, but you didn't dwell on the war itself. To an extent, all the changes around us – the building of the arsenal, the uniformed men everywhere, all the new people coming into the area to work at the arsenal – it all made life more exciting. One minute this was a place with no one around, then there were all these new people to meet.

In January 1942, I was 17-and-a-half. Until then, I'd been living-in at my job in the village. They were a family of farmers and other members of their family suddenly started coming to live with them. They wanted me to look after them too but I got fed up and just came home to stay with my mother. Then, one day at home my sister Mary told me: 'They're taking on youngsters at the arsenal, Betty.' And my ears pricked up instantly. I knew Mary had money to spend, and I didn't. I knew her money was quite good – about £3 a week – which was a lot more than I was getting. I had so little from the village job, I'd walk home from the village to save thruppence to put towards things I had to save for, like clothes.

Clothes were very important to me, even though I didn't have any money. My mother always used to say: 'Always keep something tidy to wear because if you're invited somewhere, you might be able to afford the fare there, but you won't have enough to buy something new.' At the time, my mother had bought a new costume for me on the understanding that I'd pay her back for it. I'd spotted it in a shop window in Bridgend. Her motto was: 'You can't have anything unless you work to pay for it.'

That suit was fawn with a brown pinstripe skirt and jacket. It cost a guinea [£1 1s], which was a lot then; I'd repay her a shilling a week. I'd knitted a brown jumper to go with it, but it would be a struggle to save up for brown shoes to match. So when I heard about the jobs at the arsenal, I didn't think twice. I'd join all the other under-18s they needed there. The place was divided up into different sections making ammunition, pellets, pyrotechnics, detonators, and so on, but only over-18s could work there; you were not allowed to come into contact with explosives until you'd reached 18. So they hired lots of youngsters like me to work in the 'clean' sections, like one of the canteens or in textiles. I was told I'd be working in textiles. What I didn't know was what I'd be doing was making factory clothing, overalls and headgear.

On my first day, I walked into this huge area with hundreds of people doing all sorts of different things – sitting at cutting tables, with row after row of sewing machines or pressers [to press finished goods]. Half of us had never seen a sewing machine in our lives. I had tried asking girls I knew 'what are you doing?' before I started but they all said they didn't know! They just allocated you the job and you went where they told you to go – and from that point you more or less got on with it.

That first day, they sat me down at a sewing machine. It was a heavy-duty electric machine and I was given a piece of cloth about a yard wide, two yards long. I sat there, staring at it. A woman leaned over and said: 'Just put it under.' So I did what I was told, put it under, put my foot on the treadle, and tried my best to work it. But really, I didn't have a clue. I'd gone right across the material. It scared me to death, it was so quick the way it just kept sewing, making stitches.

There was no one around to tell you off. Some of the girls working near me had already worked in a factory before, and they had a good laugh at my efforts. 'Oh, you'll get used to it,' one girl said. And sure enough, after a couple of hours at it, I learned how to control the machine. I could start making things.

We were making the uniforms for the over-18s working in the different sections – coats, belts, caps, turbans, jackets, waistcoats; trousers for the men. White coats for the women. The belts were in different colours for different shifts for the workers in the ammo section. Blue for one shift, green or red for another. When I started work there, I already wore trousers. You couldn't buy them, I just wore my brothers' cast-offs.

I worked two shifts, either day or afternoon for about eight months. It wasn't a very long journey to work by bus, because we were living less than five miles away from the arsenal. Because of the bus timetable, I tended to have a break between the time I finished work at 3.30pm and when the bus left for home. So I soon got into the routine of it.

At one point in textiles, they gave us parachutes to repair.

These were made of fine cotton, lawn, ecru-like silk; they would bring them in for us to mend, take a panel out and put a new one in. Some of the girls would take a panel out of a parachute to repair it and if the bits that were left over were any good, they'd hand them around: 'Ooh, I'll have that bit!' One girl was so good at making things, she could literally just look at you and go: 'Oh, you're 36-24-38' and then she'd cut it out, without a pattern, and make something to fit you out of parachute silk or cotton. She'd make you a petticoat or cami-knickers.

Of course, you'd have to put them on and wear them before you went home. You were always searched, every time you went in or left the arsenal. The police and the searchers were always there, at the kiosks, on the way in and out. You had your pass to show on the way in. But they could always do a random search. I was lucky; I never got caught with the parachute camiknickers. If I had, I'd probably have been out of a job, but the reason we did it was the rationing: you couldn't buy anything without the rationing coupons we all used. I know a few people got sacked for stealing powder bags [bags made by the workers in textiles to store the powder for ammunition].

I was 18 and two months when I was moved into the pellets section. Most of the girls who'd started working in textiles were moved into pellets; it was a way of easing you in. At 18, of course, the money went up; one of the big reasons the girls round here went into the arsenal was the chance to earn much better money than you got in a shop or in service. I don't care what anyone says, the main reason Bridgend was so popular with the people working there was the money. I knew of a few girls working in the munitions section of the arsenal who earned more than their fathers. They didn't dare tell them, of course.

It was dangerous work, though we didn't know it. The first work I did in pellets was to paste and wrap the pellets. You were given a piece of paper with a fluted edge both ways, a pot of paste and a paintbrush. At first, it was a round pellet, about an inch long. It was a bit like wrapping sweets; you pasted first and wrapped each pellet in the paper. Your hands got really mucky. The stuff came off in your hands as you wrapped it. You'd get bored out of your mind doing this, of course, but what you didn't realise at first was that at a certain point your hands would be yellow by the end of your shift.

You had to wear the same caps and uniforms you'd been making in textiles: a white coat, a belt and a white cap with a red band in front to indicate your shift. It was mucky, boring and repetitive work. But moving into pellets meant more money. You started work at 2pm and you'd finish about 9.30pm. If you were lucky, you got a half-hour break in the enormous canteen, a cup of tea, some chips or a sandwich.

It was noisy too. Tannoys were always going. Sometimes you'd start singing, as a means of keeping the boredom at bay. But it wasn't long before I asked if I could be moved. You can only do something like that for so long, pasting pellet after pellet and winding up with yellow on your hands all the time. So they moved me – to the stemming shop. Here, the job was to fill up rubbery things with powder from a container. We had no idea what they were used for; you did these things but you had no real inkling what they meant because you didn't ever see the end product. The powder went from a container into a box. Then it went into a large machine, with a drum at the top. My job was to pour the powder into the drum, pull the handle and then the powder came through a funnel into this rubbery thing. I think this was something that was needed to light up something else to make an explosion. It wasn't an explosive material as such, but of course, the powder was in the air all around you, so when you finished your shift, the powder had stuck to you.

And you were yellow. You were foolish if you didn't get that turban on first. In those days, we nearly all had long hair. You'd put your hair up in a round ring stocking top, tuck your hair into it and it helped keep the curls going. Then you could put the turban on over it, and over that came your cap. If it was possible, you'd put another turban on over that to protect you from 'The Yellow'.

If the powdery stuff got into your hair, it would change colour. Even if you had a little bit of hair showing, that was it, it became discoloured. If you were blonde or ginger, it went green. Black hair went red. You tried your best not to let it happen but there were still times when you'd get a little bit on your face.

The problem with the yellow was, whatever you did, it wouldn't come off. While you worked with it, you'd come home yellow. It would go through your clothes and onto your body. During the night, if you perspired, you'd find the yellow all over the sheet. It was so bad, you'd think you had jaundice. Away from the arsenal, you could always spot someone who worked in pellets. They used to call us 'Canaries'.

My mother could never get the sheets white. She always hung our sheets behind the house; she wouldn't hang any of those sheets for the neighbours to see, or anybody walking by on the road. Everyone had their own tip or bit of advice on how to get rid of the yellow, though it never seemed to work. One idea was to soak a little bit of bread in milk and try to dab it off with that when you got home. If you were daft enough to believe that, you'd try it – once. Or someone at work would come in and say: 'Cold tea gets it off.' So you'd go home and wipe cold tea over your face. It didn't work but it made you feel better to try. They did give you cream to rub into your hands and face; it was in a blue pot. I think it was Evening in Paris cream.

Going out to a dance could be a problem because of the yellow, not just being called a canary but because in the electric light at the dance hall, it just looked like you were bronze, sunburnt. One night, at a dance, one of the boys said: 'Here she comes, she's alright in this light. But you want to see her in daylight.' Then another guy in the same crowd just hit him. I watched him sliding down the wall.

But my attitude was: there was a war on, and this was our way of helping win it. I now had one brother in the Army and one in the Air Force. If we didn't do this job, who was going to do it? They had to have things to fight with, didn't they?

There was only one way out of pellets, and that was to tell the doctor at the arsenal that you had a rash. So you'd tell him: 'I've got a rash, I want to go off the section.' Most of the time, all you'd get back was: 'Oh you can't have a rash with this powder. Go back to work.' That would get your gander up. 'Get your glasses, look at this,' you'd say. Then, if you were lucky, he'd sign you off the pellets section. But you couldn't keep making a fuss all the time; then you'd get known for it. You could only do it once.

There was one doctor at the arsenal; we called him 'The Butcher'. He always had spots of blood on his coat. Years afterwards, I ran into him and he recognised me. 'Oh hello, Betty,' he said. 'Oh hello, you were the butcher,' I said. I took great delight in telling him that.

Being yellow didn't stop the men from coming on to you. There were mainly women working in my section at the arsenal but there were a few men, mostly men who had come out of working in the collieries because they had chest problems. They weren't physically fit or desirable but they'd still come on to us girls. I suppose with so many women around them, they couldn't help but try.

They weren't very romantic about it. They'd come up to you and say: 'Are you someone's budgie?' which really meant 'Are you sleeping with anyone?' Or they'd offer to help walk you to the toilet – the place was so huge, there were lots of doors to get through to get to the toilet. Then they'd sneak in the budgie question at the same time. I'd make it very clear where I stood. 'Do you think for a minute I want a reject in this place when I've got lots of handsome men up at the aerodrome? No way!'

It was true in a sense. If you were working on a night shift you couldn't do much at night but when we were on days, my sister and I would go out every night: pictures, first house, then to a dance afterwards. We'd even go to Porthcawl to a dance and then walk home. It was a five-mile walk but we didn't mind.

You'd dance with the boys in uniform – Army, Navy, Airforce boys. Sometimes when I was on day shift I'd be walking home from work on my own and you'd find boys on the road who were totally lost; you'd have to guide them to the turning for their camp. Then they'd turn round and shout: 'Are you safe, miss?' They wouldn't move until you shouted back yes, you were ok.

But there was no hanky panky. To me, hanky panky was out of the question. I had my reasons to keep myself as I wanted to be. I went out with one Airforce lad; he'd come to the house for a cuppa and all of a sudden… he was gone. You couldn't make any plans with anyone in that situation, what with people coming and going all the time. Some of the girls I knew were free and easy but at the end of the day, the men would take what they could get – leaving the girl with the consequences.

There were some Americans in the area, of course. I met some nice ones. But that was all there was to it. Sometimes the first thing a guy in uniform would say was: 'I'm married, this is a photo of my little girl,' so you could be friends, sit down and have a chat with them. And they'd want to talk about their family back home.

I liked the friendship of it, rather than the courtship; I was always more for men as friends. I did go out with one American, a tall handsome sergeant in charge of transport. He'd send me and my sister a van or a lorry to collect us from work if he couldn't make it.

At home, we were very lucky with rations and food. My mother would barter, mostly with eggs or vegetables, if she wanted some sugar or butter. There was a lot of that through the war. If there was any meat available, the men had it. They were doing hard, manual work, after all. We had our vegetables in the garden, so you could pop them into your sandwiches – peas, beans, carrots, potatoes – or you could swap them with someone who wanted to swap a bit of ham.

At one point, they moved me from stemming. A whole lot of us were moved to actually making more pellets – more yellow. Then, we were moved to making pellets out of different explosive powders – but no yellow. We'd have to work a press and there was a thick wall between the press and us. That way, if anything exploded, there was protection for us. But luckily, we never had any accidents.

The arsenal itself was never bombed, but Kenfig Hill, my area, a few miles away, was. It happened during the day and I was at work. My dad and his gang were working on the railway line when they were machine-gunned from the air. Luckily, no one was hit. But my dad picked up a bullet and brought it home, much to my mum's annoyance.

One time, the Americans organised a huge 'do', a social 'thank you' to everyone. The dance was such a big deal; I even went to the hairdressers after my day shift. I had it hanging down in ringlets, rather than the usual swept-up style. When I got to the social, everyone ignored me. So I ran to the loo, combed it out and when I came back, everyone clapped and cheered. So I didn't bother with the hairdressers again.

But at these dances there was never a chance for any of us to really talk about what we were doing. You'd just dance and enjoy the evening. But suddenly, after that social, all the Americans had gone from the area. Where we lived, no one knew about D-Day, the Normandy landings. Maybe people on the coast had some idea – they must have – but all we knew was, our American friends had gone. It meant something big was happening but it wasn't until the D-Day news came through that we realised just how big.

At work, they moved me again to the cordite section. That

was very different. I'd describe it as like playing around with a big bar of soap. We had to weigh it and then cut it to a certain length. You would have to put a protective strip on top before the cordite went into a metal container of a rocket; you couldn't damage the cordite.

When they moved you to another section, you wouldn't know a soul. The Bridgend arsenal was so huge; you'd have to start all over again with the new girls you'd be working with. One vivid memory is walking into the cordite section canteen for the first time. There must have been a hundred people sitting there. Not a familiar face in sight; a bit nerve-wracking.

'Over here, Stormy,' shouted one of the men.

I didn't know him, but 'Stormy' was my nickname at the arsenal (because there were five girls named Betty in my section) and he was obviously taking the opportunity to chat me up.

'No thanks,' I said to myself.

'I can cope with sitting on my own.'

And sure enough, I spotted a girl I did know at a table.

They must have thought I was a good worker because towards the last year of the war, I ended up in another section, making the rockets they fired from the Typhoon aircraft. Some of this was a little bit daunting. To make the rockets, you went into a hole in the floor and you were more or less buried in the hole. Only the top of your head was showing, just level with the floor. In front of you on the floor was a big umbrella type of construction, like the ribs of a big umbrella. The rockets were put on each 'rib' and our job was to fit the nose cones. Apparently, the men were too big to actually go into these holes, so it was only the girls that could do the job.

By then, of course, I had a different perspective on it all. With the various jobs I'd had handling powder, I didn't know who was going to use them. But with these rockets, you knew they were destined for the RAF to use. Like a lot of the girls, I had brothers in the Forces. So I'd think: 'Well, at least they've got the ammunition to fight.'

Of course, you had strong feelings about it all. You couldn't help it. On the bus home, the bus would go along the Bridgend by-pass and by the side of the road there was a prisoner-of-war camp; it was called Island Farm. In March 1945, 70 German prisoners-of-war had tunnelled their way out of the camp. At one stage, there were high-ranking German officers there; they still had their uniforms, their jackboots, even their medals on.

By then, of course, the war was drawing to a close, but we'd seen the state of our boys starting to come back from the prisoner-of-war camps. It really annoyed me to see those German officers still in their uniforms. They still had their batmen to help them, while our boys were in ribbons. They took away our boys' identity. Yet we didn't do it to *them*.

I could never accept that, for some reason. It just seemed so wrong. Why the hell were they allowed to keep everything, while our boys could not? I'm not a vindictive woman by any stretch of the imagination, but that one thing made me so angry. Why? My mother would always say 'The war's not over until the boys are home', so it didn't register with us that everything was going to finish so soon. Jack was in the Middle East in the Army for five years, Bob was in the Airforce; little Joe was too young to fight.

For us, at the arsenal, when it was over it was over. The war in Europe had finished, we'd had VE Day and one day at work we were told: 'Don't come back tomorrow, you're finished. Here's your pay.' It was a bit of a shock because we'd all wound up in textiles, tidying up in those last few weeks. Now they were handing us a note that said: 'You're unemployed now'. It did feel like you were just being cast aside. And you didn't know what would happen after that, you'd got so used to having regular work. All you could do was hope that something better would come along.

At first, you'd just go to the dole office once a week, to sign on for your 21 shillings. Then, one day, I turned up as usual and they handed me a green card, saying 'Go to the Don', a hall where they'd started up a new factory in Kenfig Hill. 'What was involved?' I asked.

'Oh, you work there 8 till 6pm with an hour for lunch. The pay is 18 shillings a week.'

'I'm sorry,' I said. 'I'm not doing that. Eighteen shillings a week for a job when you get 21 shillings from the dole.'

'Look, you can't refuse,' the woman at the dole office said.

'Try me,' was my retort. Again, it didn't make sense.

The arsenal itself had stopped production but other firms there took over the sewing machines. So I wound up going back to the arsenal as a machinist in 1947. I'd applied to be a machinist but they decided they wanted me to work as an examiner, checking other people's work, making raincoats, cycle capes, sou'westers, wet weather clothes. I worked there for a year.

There were things I missed about my wartime work, despite 'the yellow'. The friendships made the difference, really. During the war, you had targets. There was no money for us for meeting that target. You'd have the odd day when everything went wrong, so the other girls that worked with you on the bench would make up your target for you. In that respect, it was very friendly. You stuck together and helped each other out if you could.

In 1948, I left work completely. My mother was ill; she had terrible shingles from waist to knee. She couldn't sit or stand. I'd wind up painting her skin for her, using a feather with a purple thick liquid; this had to be painted onto her skin every day to dry up the sores because they started off as open blisters.

A year later, my dad died. He was 64. He planned to retire that December. He had a sudden haemorrhage. I'd been to the hospital to see him and went to my sister-in-law's at Port Talbot to wash and change. He died while I was there. My mother went to pieces. I couldn't leave her for any length of time, so I stayed at home, looking after her. Joe was at home too, at first; then he married and left home.

In August 1953, I went for a family day trip to Porthcawl. On the way back, a group of us, including little Leonard, our 'evacuee', now 14, went for a drink. A man started chatting to my sister Nancy. It turned out he knew her from schooldays. His name was Ivor. He lived in Kenfig Hill, the other side of the railway line from us. The next day, I popped out for some shopping and when I came back my mother said: 'There was someone looking for you, name of Ivor.'

Ivor was 35, seven years older than me; he'd been widowed after the war. I didn't know any of this until about three weeks after he'd come to our house. We literally bumped into each other in the village.

'Ooh, I've been looking for you,' he said.

My attitude, it has to be said, was a bit 'take it or leave it' but he insisted we go to the pictures that same night. And that, for us, was the beginning. We married in March 1954 and we were together for over 50 years until he died, age 87. He had dementia and I looked after him myself for nearly six years. I did work again, as a home carer, for about 20 years while Ivor was alive. And I enjoyed that too.

When he went, it was either give up or go on. And I did, somehow, go on. A lot of my old friends had died by then, so I got out and about again and made new friends. We didn't have children. There was no clear reason; it was just one of those things. I have nieces and nephews – Joe had five children – and if I pick up the phone and say 'come', they're here. We're still a close family.

At one point I got involved in trying to get a memorial set up in Bridgend for the people that died there in the war; eventually the Civic Society took it over. I had an invitation in 2012 to go to the Memorial Sunday wreath-laying at the Cenotaph in London. Of course, my family wanted me to go. But when I thought about it, I realised it might knock me over emotionally. It was the first time munitions workers were being recognised for what they did in the war.

But it had hung around for so long. These days, I get upset even at the local war memorial. The boys from the village that were killed in the war were our friends. Yet when I stop to look at it all, I enjoyed my years at Bridgend. We were young and we shared everything, good or bad. We were not rich, by any means.

But in life itself, we were rich.

CHAPTER 4

MARGARET'S STORY: THE PROCESS WORKER IN 'A' SECTION

'ONE SPARK AND YOU COULD BE DEAD'

Margaret Curtis was born in 1922 in Lanarkshire, Scotland, and went into service as a parlour maid at age 14. She was employed as a process worker at ROF Bishopton, in Strathclyde, a huge explosives factory just outside Glasgow, for three-and-a-half years. After the war, she married and moved south to live near Braintree, in Essex. Her husband, Jack, died in 2000. She has one son, two grandchildren and one great grandchild, born in 2013. This is her story:

I was a twin, but my sister died just a few hours after me. I was born at High Blantyre, Lanarkshire. They named me Margaret Reid Williams after my grandfather who was a coal miner: in the 1800s Blantyre had been a big coal-mining area. My mother's father, John Boyd, was a first-class shoemaker; he died sometime during the First World War.

71

My dad, Donald Stewart Reid, worked on the roads. He was a great dad; he loved his home and his garden. He'd been in the Army for a short time in France in WW1, just before the Armistice was signed. Being brought up in the country in a little two-storey house made for a happy childhood. We were always very well-fed, even though my dad had several different jobs. My mum just stayed at home with me until the war started and then we both wound up working at Bishopton. My mum was a wonderful knitter. You name it, she could knit it. Or cook or bake it.

As a child, you made your own enjoyment with simple things. I think we were a lot happier than children are today, to be honest. My few toys were a doll and a pram. I spent a lot of time playing with my toy sweetshop, or playing hopscotch or five stones [a very old game played with five small stones or pebbles, where one stone is tossed in the air and another is picked up before catching the first stone before it hits the ground].

You got a penny a week pocket money if you were lucky. I remember taking empty jam jars back to the Co-op; they'd give you a penny for a 2lb jar and a halfpenny for a 1lb jar. An empty lemonade bottle got you a penny. And, of course, those pennies bought you a bag of sweets.

I was quite a good child; my mum never had to tell me off very often. I wouldn't upset her for the world. Where we lived was very countrified. There was a coal mine just up the road: Blantyre in the late 1800s was a very big mining community with three mines. They had a huge disaster at Dixons Pit in 1877, Scotland's worst mining disaster when 207 miners were killed in an explosion.

Everywhere you looked, as a kid, there were 'bings' – big

piles of waste material from the mining process. You'd often see people raking amongst it to find coal. As a kid, you'd have great fun sliding down the bing on a big coal shovel. We were at the bottom of a hill and a bit further up there was a railway, which took the coal from the pit. And further up, you were right out in the countryside.

The only time I remember my dad being out of work was in the General Strike in 1926. Yet we never went hungry. And we were never cold – though you never had a day off from school when it snowed in those days. There was no electricity; we had oil lamps. The paraffin lamp hung on a wall bracket. We never had a cooker either, everything was cooked on a kitchen range; it shone from end to end. And we'd have what is called a 'rag rug' on the floor.

Mum and Dad used to make them out of strips of rag knotted through a mesh, or an old sack. Sometimes Mum would dye the strips of rag different colours. You got your milk delivered by a man with a horse and cart. I can still see him measuring it out of a churn into Mum's big jug. We brought fresh farm butter too. We did have cold running water; you heated it in the little kitchen at the back where we had our copper. That was where you boiled everything, including the clothes you washed. You just lit the fire underneath it.

One of my most vivid memories of childhood is my mum standing outside the back door doing her washing in a big wooden tub on a hefty wooden stand, complete with washboard. She'd let me turn the mangle – no spin dryers around then! At one time we moved to a house in Dundee in a place called Lochee with a small back garden. Dundee's a lovely city, but we had a robbery there. They stole money from the gas meter, and they smashed my beautiful doll to bits by sitting on it.

As I child, I had diphtheria, which was quite serious then and I wound up in hospital for six weeks. It was complete isolation; I didn't even see my parents. Then I went home for a month and I got measles: another month in hospital in isolation. As a result, I developed a real phobia of injections. To this day, I hate them.

I went to a convent school in Dundee for a while, then to a local primary and from there I went to Calder Street Secondary School in Lower Blantyre. You had to walk to school but if it rained, you were given tuppence for the bus. Sometimes I'd walk anyway and spend the money on sweets. I wasn't very clever. I quite liked History, and Music. One big childhood memory is singing 'Flow Gently Sweet Afton Among Thy Green Braes' on Burns Night.

At secondary school I always did well at cookery and laundry. My needlework was good too. Domestic Science was a big part of our secondary education – and Scottish Country Dancing, which I liked. You were taught how to clean and polish in a school bungalow – for me, that was a real relief from written or desk work. I didn't take any exams, so I left school at 14 in 1936. I'd have loved to take a course in cookery but I didn't have enough confidence in myself to do so. I was always good at the practical in cookery, but not the theory.

The only jobs around for girls like me were in domestic service. You went to the local labour exchange if you wanted to find work. And I was lucky to find a job as a parlour maid in a big house in Cambuslang, a suburb of Glasgow. I'd get a bus and then walk up the hill. I used to have to wear a black dress, a little white apron, a little white cap and black lace-up shoes and stockings.

The parlour maid's job was to help with the meals: first, serve the meals in the dining room then tidy up afterwards. I'd start work at 8.30am and get away by about 6pm. I considered myself really lucky earning about 20 shillings a month. I'd give my mum half and keep the rest. The people I worked with were a lady and gentleman with two grown-up daughters living at home. One of the daughters used to make lampshades, the other was some sort of therapist. You'd get Sundays off and one half-day off on a Saturday, unless there was a function in the house, like people coming for a meal; then you had to work all day. I was their only servant. The lady of the house did all the cooking.

Just before war broke out in 1939, we moved to a house in Springwall, Blantyre, a nice house with a garden where Dad could grow his veggies and keep chickens, so we always had nice new-laid eggs. Blantyre had a railway station too, so I'd often go to Glasgow with my mum – just a bit of window shopping and some lunch. Or my friend Mary and I would go to the pictures in Hamilton; you could get the bus from Blantyre to Hamilton.

In those days, if you saw someone wearing the same thing as you, you wouldn't like it, so I used to go all over the shops in Glasgow to find something a little bit different. My mum would knit all my jumpers and cardigans, but I still loved looking for something new: one dress has always stuck in my mind, a lovely green dress with a V yoke and a V at the back with a flared skirt. The material was some sort of knitted silk; you probably can't buy that now. But I still remember it. And the little ankle-strap black patent shoes I wore with it. I thought I was Queen of the May with those shoes.

In March 1939, my sister May was born, so Mum and I were preoccupied with the new baby. I was at home, with Mum and Dad, just three months away from my 17th birthday, when we heard the news that war had broken out. We knew it was coming; everyone knew. I did find it all a bit frightening, especially when we had the blackout, but we got used to it. Sometimes my friend Mary and I would be walking home from the pictures and the siren would go off: we'd run like mad to get home and into the Anderson shelter in our garden, waiting for the all-clear. I carried on working at the house in Cambuslang through 1940 and 1941 but it was obvious I was going to have to do something. But what? I didn't fancy the idea of the Forces; nor did my parents. In the end, my dad said: 'Don't wait for call-up, Margaret, I don't mind if you go into munitions.'

At home, we'd already had a change: my grandmother, who'd worked as a spinner in a jute mill all her life, fell over and broke her leg. So my mother brought her home to us and she stayed with us, helping look after May, the baby. So me doing war work would be ok.

In the early months of 1942, I was off to the local labour exchange to register for munitions work. I went with my friend Mary and they told us we'd be earning £2 a week, a lot more than I'd earned in service. We'd already heard it was good money, because the word had got around. And we knew it was an all-girls' environment. They hired us on the spot, but there was not much information about what work we'd be doing. They said: 'You'll be shown what to do once you start.' They did say we'd either be working in the cordite section or the gun cotton section.

On my first day I was told I'd be in gun cotton, and my friend, Mary Paterson, was in cordite. To be honest, I didn't mind what work I did. Being so close to Glasgow, a lot of the girls I'd be working with were city, salt of the earth types, who turned out to be great women to work with. And, right from the start, everyone was of the same mind: 'This is the war effort and we're all helping. That's it.'

The shift pattern was 6am to 2pm, 2pm to10pm, or 10pm to 6am. You changed your shift every week. The journey to the factory was, walk to the train station at Lower Blantyre, get the train to Bishopton and then walk to the factory gate. They'd lay on special trains for the munitions girls, so for the early shift I'd get up at 4.30am for the 5.10am train.

Once you got there you went into a special area to change; I had to wear a white jacket and trousers, a white turban and rubber boots or Wellingtons. You wore those summer and winter in the plant, but you could not leave the plant in your wellies – that was forbidden. In my section, the floors were always damp. And to prevent explosions the plant was not allowed to dry out.

There were lots of other things you were forbidden to take into the building. No metal anywhere, no safety pins, no hairpins, no matches, no ciggies – the tiniest spark could put everyone at risk from explosion. There were men, the 'danger building men', whose job it was to carry out spot checks for any dangerous items, going round all the time, double checking on us.

My section was one of a number of buildings on the site. The cordite section was extra dangerous because they were working with highly explosive nitro glycerine, so these sections were underground buildings. You could only see them by the huge grass mounds. The women who worked there wore the same uniform as us, but they had to wear felt-soled shoes for safety reasons.

I started out working in a section with 12 other women and a supervisor. We made squares and plugs out of the gun cotton. The gun cotton came to the factory by rail, then it was stored in vats to be mixed with water and other chemicals. It looked just like snowflakes. First it was processed, tested and then drained off. Then it was piped into individual machines where it arrived to be blended, or rather shovelled, from one bin to another and back again until it was ready for use.

Our job was to get it into the squares or plugs before it was pressed into shape. There was no music or radio to listen to while you were working. You had to concentrate completely on what you were doing because you were working processing machines. One of the supervisors showed me how to work the machinery on my first day. I soon got the hang of it.

The gun cotton would be handed to me in a small container; it was a fine powder, not quite the consistency of flour but near enough. I'd put a funnel into the machine and shake the stuff in until it was filled. Then I'd close the lid on the machine, push the handle to press it and the gun cotton would come out in an oblong shape as a block, about two inches deep.

Then these blocks would go into a tray to be carried over to a pressing house where it was weighed and pressed into shape. Some of the blocks would make little round shapes and these too went over to the pressing house. After that, in the drilling room, holes would be put into the blocks and the cores and then they were dipped in acetone, in another section, for protective coating. Then these were dispatched and became part of the detonator assembly for depth charges and other explosive devices.

We were all told we had set targets to achieve on each shift. Whatever the target was, it had to be achieved on that shift. Sometimes the forewoman would come up and ask you to do 'a bit extra'. You just had to get it done. God help you if you didn't keep the girl in the pressing house going.

I'll never forget Annie, the girl in the pressing house; she was a cracking worker, a big girl and an absolute gem. She'd stand there, hands on her hips waiting for us to fill up the trays. 'WHAT THE BLOODY HELL ARE YOU DOING?' she'd yell at us. At first, I was a bit worried about Annie's yelling. Then I got it. We all wanted to do our best. She just happened to be an extremely hard worker. In fact, she kept us on our toes.

They really didn't tell you too much. You didn't even know where these things that we were making were going. We just knew that what we were making was something related to depth charges. In fact, what we were doing was making explosives for the mines being carried by the submarines. You didn't have any real sense that it was dangerous work in the beginning. But then, after I'd only been there for a few days, something happened that surprised me. The danger man came over to me. He had spotted something I'd completely forgotten to remove – a small Kirby grip in my hair.

'Ooh sorry, forgot to take it out,' I told him. Then I got a bit of a shock. I wasn't expecting to hear what he said next: 'You've not been here long, so I'll let you off. But do it again and you'll be suspended.'

I was mortified. Then I realised just how dangerous anything metal, no matter how small, could be. One little spark and we could all be dead, every one of us, burnt to a cinder. Stories about the danger would go round all the time, as I soon discovered. One day, in the canteen, a girl told me how someone had been cleaning a big machine with a brush. Somehow, a single hair from the brush had got into the mechanism. It caused one spark and whoosh, everything went up, though the girl didn't know for sure if any workers had been killed.

Then, after a few months at Bishopton, I came in one day to start my 2 to 10pm shift. It was a lovely day. I was on my way to the changing room. For some reason, I had a library book in my hand. The next thing, I heard a huge explosion and the book flew up in the air and hit the ground. One of the underground cordite houses had just gone up. The rumour we heard later was that two or three people were killed. But even though you couldn't see a thing, it brought it all home to you: these explosives really do kill. And we're working with them; anything can happen. You never did find out why or what happened that day. They kept all those sorts of things, that information, close. We didn't work in the area, so it was nothing to do with us. You just had to get on with what you were doing.

Mum started working in the gun cotton section after me. If we were on the same shift, sometimes we'd go in the canteen together; that was the one place where we'd all let off steam. The food was nice: soup, steak pie, about a shilling for a hot meal. I don't know if they were allowed a little bit more, but it was always plentiful. No desserts, mind you, thanks to sugar rationing.

The canteen was very sociable. The men would come in from the machine shops and a big group of us would sit there, playing cards. The radio would be playing, Joe Loss, Henry Hall, Betty Driver. I'd always join in when everyone started the singing. We used to sing a lot. Your time in the canteen depended on the shift. On day shift you'd get a break at 12 noon; you'd get an hour. There was no point in leaving the factory, and anyway, you weren't allowed out until your shift ended.

There was one time I remember when the machines were shut off and we all stopped work completely. I was on night shift and a call came through, then the sirens went off. We were all told to leave and go straight to the changing room. All the lights went out around us and somehow we had to make our way to the changing rooms. It was so, so scary. Then we had to stay there, praying for the sound of the all-clear. Mum was working night shift too that night, for some reason.

Of course it was frightening, hearing the drones of planes overhead with no idea where they were heading. That particular night, I realised afterwards, the entire factory had shut down, though luckily, that night, the German planes didn't do any damage. We'd already experienced that terrible time in March 1941, the two nights when the German bombers attacked the Clyde shipyard and bombed Clydebank. I'd been at home both nights. My dad just plucked up little May from her bed and rushed us all into the Anderson shelter.

It's funny what you remember. May was only two at the time and she seemed fascinated by the reflections of lights coming through the small space under the door of the shelter. We could hear it all because the River Clyde runs through the middle of Blantyre on its way to Glasgow, so the bombers were using the area as a road map. Sure enough, in the morning, when we came out after the all-clear, we found the remains of a spent flare in the garden. [The Luftwaffe dropped flares, attached to a tiny parachute, for target illumination and marking at night.]

Dad was an air raid warden, just along the road from us. He had to go to Clydebank immediately afterwards. I'd never seen him look so shaken when he got home. Afterwards, I went with Mary to have a look. It was tragic to see those houses reduced to rubble, some of them with their sides sliced off, so all you could see were wrecked rooms, the pictures still on the walls, the furniture hanging in midair. Later we heard that a German reconnaissance plane was shot down on its way back from the raid. The story was that the plane was loaded with film. When it was developed, there were very clear pictures of the Bishopton plant. I never wanted to stop and think about what could have happened if that plane had made it back to Germany with those photos.

In a way, because you had to keep your wits about you working with the machines, having to concentrate, that helped; it stopped you thinking too much about all the terrible things that were going on. And we all helped one another there. You had a lot of work to get through, but there was always someone around to help you if you needed it. Working with a really nice bunch of girls made all the difference.

Mary – I wound up calling her 'Cordite Mary' because there were so many Marys at work – and I were young and lively; the war definitely wasn't going to stop us from going out. Strangely enough, you felt safe as houses if you did have to walk around at night. There'd be men in uniform walking back to their billets. Or the air raid wardens would spot you and tell you to hurry home. If I went to a dance near home at the Co-op Hall in Blantyre, my dad would stand and wait by the door at the dance – to walk me home.

Depending on shifts (you could get a late pass sometimes for your shift, if you had a good enough reason) we'd sometimes go dancing in Glasgow, to Green's Playhouse, dancing to Joe Loss. Or we'd go to theatre in Glasgow for variety shows at the Kings, the Alhambra or the Pavilion. Sometimes it was Mum and I going to the pictures at the Doolkit, the local Blantyre cinema: George Raft, Anna May Wong, Richard Tauber, Bing Crosby, Deanna Durbin. Oh, I liked my musicals.

The working hours didn't leave a lot of time to go looking for boyfriends, mind you. I'd had my first-ever date, a young man from Blantyre who was in the RAF, but he wasn't my type. But it was at the La Scala Picture House, in Hamilton, where I met the love of my life one night in 1942. I was 20 and it was a few months after I'd started at Bishopton. I was there with Cordite Mary, as usual. Just as we were walking towards our seats, someone whistled at us from the seats behind.

Of course I turned round and saw two young uniformed blokes grinning at us like mad. I must admit, I gave them a dirty look. But once we came out after the film, Mary nudged me. There they were again, the same two young men, smiling at us. I would have held back, but not Cordite Mary. 'Come on, let's see what they're like,' she urged me. It turned out they were both in their twenties, in the Royal Army Service Corps.

'I'm Fred and he's Jack,' one said, pointing to his friend who had the most wonderful, piercing blue eyes; you couldn't miss them. 'We're stationed in Hamilton, near the race track. Er… can we walk you home?'

I still hesitated, despite the piercing blue eyes. Then Cordite Mary gave me another nudge. 'They're alright,' she hissed at me.

She was right. When he left me at my front door, Jack had already asked me out to the pictures and I'd promptly said yes. He was my first proper boyfriend. I told Mum and she insisted on meeting him, of course. By then, I'd found out more. Jack had been called up in 1940. He was from down south, a place called Aldham in Essex. He'd trained at Felixstowe then he'd been posted up to Scotland.

I knew Jack for about three or four weeks before he was posted overseas. Much later, when I'd told him how his blue eyes were the first thing I'd noticed, he said *his* first thought was 'she's got nice legs'. But once we started courting, he didn't like me wearing short skirts! Jack ended up being posted to North Africa, then Italy. We'd manage to write. Sometimes you got the letters, sometimes you didn't but in November 1943, just before my 21st birthday, we got really bad news. Jack was missing. He was a driver. He was driving a brigadier and a major in southern Italy when they were stopped by German gunfire and captured. Somehow, one of the two men's wives had managed to get a letter to Jack's mum to tell her he'd been reported missing.

As you can imagine, I was very upset. But I kept telling myself: 'He'll be alright. I know he will.' Even when the postman turned up one day to hand me a pile of letters I'd written to Jack that were being returned to me, I still clung to the idea that 'missing' meant that. It didn't mean 'gone for good', I kept telling myself – though of course it did for many women, back then.

At work, most of the other women had husbands away in the Army, so we'd always be comparing notes, what we were hearing, whether letters had been coming through. One or two of the older ones were very reassuring.

'Look, he was with a brigadier and a major. He'll be ok, Margaret,' they told me. In fact, that was exactly the case. When the Germans captured them, one of the officers pointed to Jack and said: 'He's my batman'. [The term 'batman' means personal servant to a commissioned officer.] So he was ok.

In January 1944, my mother got good news from the major's wife. She'd written to say all three men were in Germany as PoWs. Jack was alive! I'd lived in perpetual hope – and my optimism paid off. Though years later, Jack told me that being a prisoner-of-war was so bad, if it hadn't been for the Red Cross food parcels, he didn't know what he'd have done. Those parcels, distributed by the Red Cross to prison camps in Germany, were a lifeline. They didn't contain much – some tobacco, a bar of soap, a tiny packet of tea or a tin of sardines – but they made the difference.

So those last 18 months of the war weren't too bad for us. Jack was alive, that was what mattered, though the winter of 1944/45 was an exceptionally harsh one. You'd often go out in the morning with socks over your shoes, to protect them, and it was a real slog getting to work, especially if you were on the early shift. But somehow you'd get there and do your shift. And in our section, we were lucky compared to the women working in cordite, which was so dangerous.

When the news came through that the war was over, my group, 'A' shift, were the last to leave. I finally finished work at the factory in October 1945. The factory itself stayed open, but we women had to pack everything up. We wore these enormous greatcoats they'd supplied us to pack up. And although everyone was relieved that it was all over, there was a bit of sadness at saying goodbye to the girls. In spite of it all, we'd had some happy times together, and that helped get us through it all.

That spring, I'd got the best news of all from Jack's mother: he was on his way home. He'd already asked me to marry him in one of his letters. I'd written back and said: 'As long as you don't think you're making a mistake, Jack.' Sometimes I'd get a couple of letters and then a few months would go by and nothing… so when I heard the news he'd be back, I was really over the moon.

Oh, how happy we were to see each other again that day in April! Jack wasn't very well. He had a skin disorder and other health problems. But we were officially engaged in May 1945 and my parents were very happy about it all. My gran liked him so much; she would never hear a word said against him!

On 22 December 1945, not long after I'd turned 23, we got married in the church in Blantyre. My little sister May was bridesmaid. It was a bitterly cold day. I had to go to Edinburgh to get the wedding dress because they were so hard to come by. Lots of girls I knew had to borrow wedding dresses because they were so difficult to find. But I found a lovely long white slipper satin dress with a white silk veil. I even had a bouquet with pink and white carnations.

We had about a hundred guests at the Community Hall in Blantyre. It was a smashing party, all the girls I'd worked with on the A shift were there and I managed to have a two-tier wedding cake. We had a week's honeymoon at Clacton-on-Sea in Essex. Jack's Aunty Mary owned a guest house, so we stayed there. Jack was a butcher by trade, but he was out of a job. He found work as a mechanic up in Scotland for a year or so, but although it was a bit of a wrench for me, we decided to move down to Colchester in 1947.

At first, we had to live with Jack's parents, then he got a job working on a farm in Stanway, and there was a little cottage we could live in that came with the job, just across the road from where I live now. It had a nice garden, so we could grow our own vegetables. We even had a few chickens – and a lovely dog called Judy.

My son, Donald, was born in 1949 and three years later I went to work in a children's home in Stanway. One of my friends looked after Donald in the day and during the school holidays they let me bring him with me. When Donald started primary school, I started working in a pub in Colchester. I started off cleaning and wound up as their cook for many years. Then I got a job in a local secondary school kitchen. Many years on, Donald taught there; Jack was an assistant caretaker and I was in the kitchen. I carried on working there until I retired.

Jack had been in pretty good health until his late seventies but he had a series of mini strokes and went downhill in the last couple of years of his life. He died in his sleep at

Colchester Hospital in March 2000, age 81. We'd had a wonderful, happy marriage for 55 years.

On the whole, when I look back to that time in munitions, I think we did a good job. What I could never understand was later on, you'd hear about the Land Girls or the girls that worked in aircraft factories but not very much about the girls who did the job I did. Where's our medal? I do think we should have a bit of recognition. We were young, of course. So we did have the stamina for 12-hour working days – by the time you got home from your shift, that's what you'd done.

But the work had to be done. If you didn't do the work, you'd have been thrown out, no question. If you turned up five minutes late, you lost 15 minutes pay. You might have got away with it if your train had been late for some reason. But that didn't happen very often. Was it fair treatment? I think it was the same for everyone. You had to know your job. You had to be 100 per cent responsible. And you had to be prepared to pitch in. The supervisor only had to say, 'Oh we need extra' and those girls would work like hell to get it done and out. There was no overtime in A Section – everyone just pitched in and worked that wee bit harder.

Have we lost that sense of being there for each other, helping each other out? I don't think people are as dedicated to looking out for one other as they were then. There we all were, walking around there in the dark, in the middle of a war, bombs falling from the skies, lives upside down all the time. Nowadays – well, people are too scared to cross the road. You had neighbours then who'd run into your home and help you, whatever happened. Nowadays you might not even know who your next door neighbour is.

CHAPTER 5

IVY'S STORY: THE GIRL WITH THE LATHE

'SHE SCREAMED THE PLACE DOWN: SHE WAS COMPLETELY SCALPED'

Ivy Gardiner was born in Wallasey, Cheshire, in 1924. At 15, she went to work as a factory hand at Lever Brothers (now known as Unilever) at its Port Sunlight village complex in the Wirral, Liverpool. When production at Lever Brothers switched to munitions in 1940, she assembled jeeps and worked as a lathe turner, making undercarriages for bombers, until war ended. Widowed at age 52, after 29 years of marriage to her childhood sweetheart, Wilf, she has one daughter and two grandchildren. In 2012, Ivy was awarded the MBE (Member of the British Empire) for dedicating nearly nine decades of her life to the Brownie movement, which she joined as a five-year-old in 1929. This is her story:

Wilf and I were married just after VE Day. As a wedding present, my mother paid for us to fly to the Isle of Man – on one of the first planes to leave Liverpool airport after the war. It was our first-ever flight, and when I looked out and saw all the guns along Liverpool Bay from above, it was a sharp reminder of what we'd all lived through.

'Oh Wilf, thank heavens it's all over,' I said to my new husband, who was holding my hand tightly as we went up. Wilf just looked at me and smiled. And then I remembered that night when the German bomb came through the roof of the Ritz Cinema in Birkenhead and killed 10 people. The bomb had exploded right in front of the circle seats. Wilf and I had been there, in the circle, with all the other courting couples. But we'd ducked out, just in time.

The air raid siren went off and I'd leapt out of my seat immediately – with Wilf somewhat reluctantly trailing out after me. Now, I didn't even have to mention that terrible night to Wilf. He knew exactly what I was thinking. 'Oh, Ivy, we've been so lucky,' he sighed. 'Suppose I'd been daft enough not to follow you.'

Things like that you never ever forget. We'd all been desperate for war to be over. But the memory of it all, well, when you talk about it now, you think 'did we really get through all that?' Yet when you were doing it, you never thought about it at all. Not really.

My dad, Albert Reston, was in the Navy in the First World War and worked for a time at Cammell Lairds, Birkenhead, as a boiler maker. One of the ships he helped build came in recently into the Mersey. But after the war, he couldn't get work. It was tough for us. My grandma had a shop, a general store and a coal yard; people would come to fill a bag of coal up for sixpence. I do remember that as a kid. I was the eldest. My brother, Ronald, came two-and-a-half years after me.

I was just nine years old when my dad died, in 1931. It was

very, very sad what happened. He got tuberculosis but not that bad, borderline. But for some reason, they chose him to participate in a drug trial in a convalescent home in Market Drayton, Cheshire. It was a place where they took people with TB. There were nine other people there, men and women. None of them were very bad.

The trial involved an injection. And my dad bled to death. My mother, Annie, never even got his death certificate until he'd been buried. She was left widowed at age 32, with two children. At the time we lived in Clawton, Birkenhead, in a terraced house next door to my grandmother. She watched over us while my mother went out to work. My mother took anything she could get. At one point she had three jobs, though she did get a widow's pension: ten shillings plus five shillings for me and three shillings for my brother. Less than £1 a week. And the rent on the little house we lived in was twelve shillings and sixpence.

She worked so hard, my mum, she was exhausted all the time. Then one day, she just collapsed on the floor in front of me. I ran down to Grandma, her mother, and she made a decision: the best thing for us was to all get a house together. The house we rented had a sitting room and a bedroom for each of us, and my mother continued to work. In those days, if you saw a house you liked the look of, and you were ok to decorate it, you would tell the landlord: 'This place needs decorating, so how many weeks' rent can I have off?'

We moved around a lot in the thirties. We lived in about five different houses. One house in Eton Road, Birkenhead, was a very nice house; that's where we were living when war broke out. The thing was, you didn't envy others because no one you knew had anything in those days. I had lots of friends. I went to every social evening the church put on. I really was a joiner. I joined the Brownies at five, the Girl Guides at 11, a long, long association. And I was a good runner, one of the best runners at my school, St Johns, a church school in Birkenhead. After that I went to Conway High. I didn't pass the 11-plus; we were moving at the time we were doing it. So I left school at 14.

At first, I didn't want a job. But then they put on a course at the college in Birkenhead to show you how to get a job, teaching you how to behave in the adult world, instead of just being a schoolgirl. One day, coming back from the course, I spotted an advert in the window of a florist shop: 'Apprentice wanted. Two shillings and sixpence a week'. I got the job. I handed my mother the two shillings and got the sixpence back from her. That was my pocket money. By then, she was working as a waitress at weddings, that sort of thing.

I stayed at the florists for a year. I did learn — I was shown how to make wreaths, but the moss you used to make them with was full of ants. They'd crawl all over me. It was horrible. In the end, I'd go to bed at night and have nightmares about the ants. My mother had been quite keen for me to be a florist, but when she saw the state I was in she said: 'Enough of this.' And anyway, I'd already had a bright idea. Why didn't I try to get an interview at Lever Brothers at Port Sunlight?

They were the biggest employers in the area and everyone knew it was a beautiful place to work, a purpose built 'village' where they really looked after the employees. If you got a job there as a youngster, you went from one department to another. They'd move you around; it was a huge place. And it only took 20 minutes on the bus to get there.

They took me. I started out working in the Lux flakes section. [Lux was a popular laundry and beauty soap at the time.] I worked on a machine filling the soap packets. The packets came along to us on a conveyor belt and we boxed them. You had to keep up with the machines. Then they sent me over to Bromborough Village, where they made Stork margarine. They gave you clogs to wear there because there was always water under your feet – because the machines were always being steamed. They told us we could keep the margarine that was past its sell-by date and couldn't be sold. I'd be scooping it out of the carton with my hands; then it went in a big vat to be sent off for making soap. My hands were soon beautifully soft. The clogs were comfortable, too! After Stork, they sent me back to Lux Flakes again. I was still under 18 and they didn't employ you fulltime until you'd turned 18. But being moved around got me used to working in a factory environment.

I was a few months off my 18th birthday when war broke out. By January 1940, just after my birthday, I knew I had to do something. Everyone kept saying the women's call-up was coming at some point. I belonged to a youth club then and, of course, most of the boys I knew were already joining up. I had the idea in my mind that I'd go into the Army. That would be my war work. So I went down to the city centre in Liverpool to get all the forms to fill in and take home.

We'd already set up the cellar in our house in case of air raids. We'd set it up for comfort: single beds, a coal fire and blackout curtains, of course. There was even an electric cooker and taps for washing. So when the bombs started coming, we were ready, organised. We thought we could live down there if we had to.

When I got back from Liverpool that evening, I went down to the cellar. My family and some friends were already down there. One of my brother Ronald's friends was in the Navy, another in the Army and my boyfriend, Wilf, was down there too, waiting for me. Wilf came from an Army family – all his brothers had gone into the Navy. But Wilf was deaf in one ear, so he wound up in a reserved occupation, as a fitter and turner for the Navy. He tried to get into the Army, he so wanted to make his contribution, like everyone else. But it wasn't to be. Everyone in the cellar wanted to know where I'd been that day.

'Oh, she's been to join the Army,' my mother told them.

'Oh no! Not our Ivy,' quipped my brother. 'She's not going to be an officer's comforter.'

And they all roared with laughter.

Of course, I knew they were kidding. But it still frightened the life out of me, an innocent teenager. If you were called up for the Army, you wound up working in the kitchen. But if you joined the Army voluntarily, you could ask where you wanted to go. That was what I'd been thinking, anyway. Now I wasn't so sure about it all.

The following Monday I went into work as usual and I was delivering letters to the foremen in each department. That was my job: take the letters round the different sections. In one section, I overheard some men talking. The printing shop was closing down, they said, and Levers was going to be doing war work. Munitions, they said. My ears immediately pricked up. I'd already decided the Army was not for me. 'I can go there,' I thought to myself. My mother was widowed; she needed my money. Most of the girls I knew who were going into munitions were office girls with husbands away in the Army. Nice girls, not rough.

I was quite a shy girl. Yet that day, for some reason, I plucked up enough courage, as I made my way round handing over the letters, to stop and ask one of the nicer foremen if he thought I could start doing munitions work. 'I'm 18 now,' I told him hesitantly. He looked at me quizzically. Back then, you didn't pipe up like that to adults, especially the bosses.

'You're a cheeky one,' he said. 'Ok, go and see Mr So and So.'

And that was how I volunteered to work in munitions. I started the following week. My first job was helping assemble jeeps. The jeeps would come into the shop unpacked, just the wheels and the undercarriage. Then we'd have to put the tyres on. Then you moved it around until the chassis came down and you put that on. Then you pushed it around again and you put the tarpaulin on.

You started at one end of the shop and at various different points you were putting things onto the jeep. Until the last stage, the greasepit. You greased underneath the jeep – you were shown exactly how to grease it all with a grease gun. Then you put the water into the tank, then the oil into the tank and then you were allowed to start it and take it outside the shop, where someone else took over and put the jeeps into rows.

In that section we were on day shift, 8am to 5pm with an hour for lunch. I was given a pair of navy blue overalls and you wore trousers underneath, which you had to buy yourself. There was a blue hat with a band on, to denote your shift, but you didn't really need a hat while you were working. There were some men also making the jeeps. But not that many, they tended to be men who were not fit enough for the Forces. There was just one older man in the store room. You hardly ever saw the foremen. Now and again you saw the managers.

It was a happy time, in a way. There was *Music While You Work* playing from a radio high up on the ceiling. And a meal for sixpence in the canteen. Wilf was making rum barrels for the Navy in Birkenhead. I was making eight shillings a week when I started out in Lux Flakes. Now I was making nearly double: 15 shillings a week on the jeeps, which was a lot. And we were still working days. The journey to work wasn't that bad: a quarter-of-a-mile walk from home and then onto a special bus to take us to Lever Brothers. Go through the main gate, show anything you were carrying, always two men standing there to check everyone.

I didn't realise at the time – it wasn't until many years later that I found out – that glycerine was stored at the factory, round the sides of the huge 'shops'. If a bomb had ever hit us, it would have set the place on fire immediately. It would have been terrible. Though at that point, there wasn't much bombing in our area.

In December 1941, the jeeps section was closed down. I was told I'd be sent to Chester for two months to learn how to be a turner at an engineering factory. Little did I know but that would turn out to be an exhausting couple of months. The training hours were either 6am to 2pm or 2pm to 10pm. Leave the house at 4.30am and walk two-and-a-half miles in the blackout to get a train to Chester for 5.45am, then run like mad through Chester Cathedral to the factory for 6am.

For this, we wore the same overalls, but with a peaked hat, where you had to tuck all your hair in, every single hair. Anything that was loose or dangling could get caught in the machinery. So your sleeves, anything like that, had to be very tight. You had to be very, very careful. If anything did get caught in the machine, that was it – unless someone was quick, saw you and managed to stop the machine.

I was 21 while I was doing the training. And because it meant doing shift work, there was more money – 25 shillings a week. They were teaching us how to use a lathe. So as we were training, we knew we'd be making the 'legs', the undercarriage for Dowty aircraft bombers, big US bombers. I was training on the lathe, another girl was training on a drilling machine, another on a milling machine. I could only operate a lathe, but there were different sizes of lathe, so there was a lot to learn. Then it was back to Lever Brothers. While we'd been away, new machines had been set up and now we had been trained to operate them.

I'm quite tiny, just 4ft 11in, and at the time I weighed about 7st 10lb. But they put me on a huge machine, a Gisholt lathe. How it worked was, I'd be given a solid piece of steel – the men had to put it into the machine for me because it was too heavy – and there was a big drill attached at the other end. I had to get the drill to the centre of the piece of steel, then you had a tap, with oil and water, and I had to direct that to the drill.

Then I could start the machine, what they called 'the feed'. Lift the handle and start to feed the drill into the steel, the oil and water ran down into a big trough. I'd have to stand in the trough to centre the drill into the steel I was drilling. As the drill turned round and bored into the machine, what we call 'swarf' came from the drill – ringlets of steel that came out of the hole you were making. You had to keep watching it all the time. Not so surprisingly, I didn't stay on this for very long.

'Who on earth put that little girl on that big machine?' someone said. It was obvious you needed a man's strength to lift the steel.

Then they set me to work on a capstan lathe. I did the same thing but with smaller pieces of steel and smaller drills. So of course, I could get through the work much faster. You needed to keep your eye on the swarf all the time, mind you. If the swarf got stuck, the drill broke. So instead of making the bomber undercarriage itself, I was now making the attachments for the undercarriage. I had to work with copper too. That didn't come out like ringlets; it would spit out. And it burnt you so you had to wear gloves.

Once, on night shift, I yawned without thinking, and a piece of copper spat out onto my tongue. Someone went and got the nurse and she came out, looked at me and laughed. 'Drink milk,' she said. It was a bit of a shock for me. It did burn. I had an 's' on my tongue for ages. I drank milk for a week. But later on, it went away. It was just a burn. Everyone around me thought it was very funny. 'That'll teach you to open your mouth, Ivy,' one of the men joked in the canteen.

But not long after that, something really dreadful happened. There were a few other girls working drilling machines in the same section as me. This particular girl worked with her back to me. 'Tuck your hair under your hat,' she was told, time and time again. But for some reason, she would never do it. On this shift, I saw her bend over to look at something, and the drill caught her hair. It scalped her. I saw it happen, right in front of my eyes. It was horrendous.

She was screaming and there was blood everywhere – tiny pinpricks where the hair root had been. She screamed the place down. It was dreadful. She was in a terrible state. The drill had yanked her hair out by the roots, so it would never grow again. She was completely scalped. They helped her, got an ambulance to take her away. But we never saw her any more after that. That frightened the life out of everyone. It was a big lesson to all of us to make sure we had completely tucked every single hair into our caps.

A couple of months after that accident, I was moved again. This time it was on to a centre lathe, which was small. This time I would be making piston rods for engines. To do this work, you had to be very precise about what you did. Luckily, I had steady hands. They were always keeping a lookout for girls with steady hands. With this type of work you really had to be concentrating very hard, watch what you did. But you were tired all the time, even when you were not working. If you were working 6am to 2pm and there'd been a bombing raid, there were no buses to get you to work. You had to walk a couple of miles before a bus caught up with you.

In the end, my grandmother got me a bike. You were given a tin hat to put on if the air raid warden suddenly blew the whistle and you were cycling to or from work. But at work, if there was an air raid, you kept working through the bombing. We were never allowed to stop work and go into the shelter. It was just non-stop. You would be on 6am to 2pm and then you changed on the weekends to 2pm to10pm on Sunday night and then back on 10am to 6pm again. There were no public holidays apart from Christmas Day; you were allowed that off, but not New Year.

You got one week's holiday a year. Mum and I went to Blackpool twice, and once we went to the Isle of Man, where we could see the internees in the hotels in Douglas with all the barbed wire around. But it was peaceful there. A week of peace and quiet away from the noise of the shop floor – and the air raid warnings.

My brother Ronald was 14 when war broke out. He was a messenger boy for the ARP [Air Raid Precautions, all volunteers] with a bike and tin hat. He'd take messages from one air raid shelter to the next, at night, through all the bombing. One day the police came to the front door. Mum went white when she saw them but they said: 'Your son Ronald's in the hospital.' He'd smashed into a concrete pillar in the Birkenhead Park. His face was in a terrible state but he was alive.

My mother was also an ARP, doing so many shifts a week. There was a big tunnel underneath Birkenhead Park where people would shelter; they had a paid air raid warden there all the time. One night, the bombs hit all the roads around the park: people in the park panicked, wanting to get out.

The warden stood there at the entrance and said: 'Anyone wants to get out, they have to get past me.' He was right. There were land mines: that was why he stopped the people from rushing out. He had a hook hand from an accident as a child. After the war, the people clubbed together and bought him an artificial hand. He was a hero.

But it wasn't all bombs and shelters. We did have fun, too. The girls I worked with used to go to tea dances in Liverpool when they had time off and of course, there was still the pictures, though they used to close pretty early. That trip to the pictures I never forgot. Wilf and I were sitting in the Ritz Cinema in Claughton Road. We'd seen most of the film. Then the sirens went: air raid coming.

'Oh, I'm going, Wilf,' I said, and jumped up straight away. For a minute, Wilf hesitated. He wanted to stay, see the end of the film. But being sensible, of course, he came out after me. It was a very clear night. As we walked along, we looked up at the sky and there were the silver planes, getting ready to bomb us. And a bit later, from the distance, you could see the bombs coming down – exactly where we'd been watching the film at the Ritz.

That was the sort of thing that happened to people. Someone made a snap decision to get out, go home or do something suddenly – and it could be all over for them. Going to work after a bombing, particularly on a bike, you'd see the houses down and the bodies – which were always round the chimney stacks. The chimney stacks were big then, so if the house was hit, people would be swept up and the chimney pots stopped them falling off the roof. You saw it all.

You'd hear the bombs, of course, if you slept in our shelter, the cellar. But you usually knew it wasn't close – unless the house shook, then you knew it really was close. We were lucky in our road; all we had were cracked windows. But not everyone was so lucky. Of the whole group of boys I knew from the youth club, most of them went into the Air Force – and were dead within a year. Just two boys out of our group came home. One lad had been in a Japanese prison camp. He died a year after he came back. The other one was torpedoed and spent a week in a lifeboat until he was rescued. He went back into the Navy.

We were so longing for peace; for us to win the war. The propaganda – 'Be Like Dad, Keep Mum' – was wonderful. Nobody ever thought we'd lose. In the factory, there were people from all over: war refugees from Poland, men from the West Indies, they were great fun. One of the girls in my section wound up marrying one of them.

In the building next to ours, there were girls packing the bullets. We were never allowed in there. And we never saw those girls on the bus or even on the same shift. The men always told us: 'You can't go in there, girls. Just stay in your own place.' After I'd started working on the jeeps, there was so much pressure to get the work done, we didn't even get time to go to the canteen to eat; the food was brought to us and we ate it in a little side room. So we never got to see those girls in the next building.

All you'd ever hear were rumours. It was all very vague. All I knew was, those girls were packing bullets and the rest of us better keep away, it was so dangerous. I don't even think they knew themselves it was dangerous. But I heard that there were really bad accidents, some of those girls didn't live very long. One friend of a friend died after an accident.

Once we all knew about D-Day, of course, that was the real turning point. We knew then that the war was going our way. I turned 21 at the end of that year. I came home after the 6am shift and my mother gave me my birthday gift: one little chocolate éclair. She'd queued an hour to get it.

By the New Year of 1945, the munitions work stopped. Ten weeks after that, I went back to normal factory work for Lever Brothers for a matter of weeks. I finished up at the factory just before Wilf and I got married on May 25th, just after VE Day. The wedding was lovely but there wasn't much

Above: 1940: King George VI examines a tracer shell at a Midlands ammunition factory as his wife, Queen Elizabeth, chats with a worker. © *Getty Images*

Below: The Royal visits to the arms factories were a huge morale booster for the workers.

© *Getty Images*

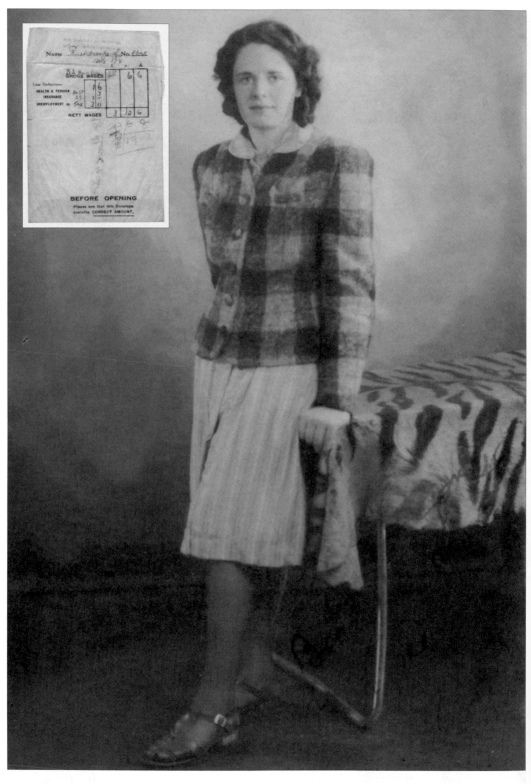

Above: Maisie Jagger: she missed her family so much that she was transferred back to Dagenham, Essex, after making gun cartridge cases in Worcester for 18 months.

Inset: Maisie's paypacket, March, 1942: £3.12s.6d for a week's work making parachutes.

Above: Ivy Gardiner's wedding day, May 1945. Like thousands of other couples, Ivy and Wilf tied the knot just after VE Day.

Above: Acton, London, 1944: Older women workers making cannon shells.   © *Getty Images*

Below Left: Betty Nettle: Betty started work at the Bridgend Arsenal as a 17 year old in 1942.

Below Right: Betty (*centre*) with two of her sisters. The Bridgend factory transformed the lives of thousands of women in the area.

Above: Margaret Proudlock (*left*) and her good friend Sadie. They worked as a pair, punching cotton into pans of hot acid.

Below: Margaret (*seated, front row, far right*) and her co-workers at Drungans: they'd often burst into song to keep themselves cheerful.

Above: An early shift at the bomb factory. © *Getty Images*

Below: Electric arc welding at a major shell producing factory. © *Getty Images*

Above: Doing her bit: working on shell caps in the busy afternoon shift. © *Getty Images*

Below: Munitions women drink milk to reduce the harmful effects of their exposure to lead.

© *Getty Images*

Above Left: Margaret Curtis and husband Jack on their wedding day, December 1945.

Above Right: Margaret's reference when she left the factory as the war ended in 1945.

Below: 1942: Margaret, (*seated*) with some of the other girls working on 'A' Shift.

to eat. Spam salad was the best you could get for a wedding reception. My mum and Wilf's mum got together to see what they could do. Even little things, like icing for the cake, were hard to find. They did a lot of swapping of coupons and got everyone they knew to hunt around for icing. And in the end, I had a three-tier cake with real icing. And I don't know how my mother managed it, but she got a quarter of a pig. Wonderful! Pork salad and a big cake with real icing. You hadn't seen things like that for ages.

All my friends got married in borrowed white dresses. In the end, I bought a blue silk coat in Liverpool, a hat with an ostrich feather with veiling. I've still got that ostrich feather. The bride and groom were both virgins on their wedding night. Neither of us had a clue what to do because no one told you or talked about it in those days.

At first, we went to live with my mother-in-law in North Birkenhead and wound up staying with her until 1950, when my daughter Lynn was born. That was when Wilf started working as a fitter at Ellesmere Port and we managed to get a council house just down the road from where I live now. Our life was as happy and contented as we'd hoped it would be over the years. In 1974, Wilf and I went on holiday in October, to Torquay. We were in a dancing group. We were going dancing almost every night. But then, out of the blue, our world began to topple. On that holiday, Wilf discovered a lump on his neck.

I wasn't too bothered at first but Wilf sensed, immediately, that something had happened to him: 'Ivy, if I've got something wrong and I'm going to die, promise you won't tell me,' he said to me one day, sitting in the park. I just nodded. A promise was a promise. But surely it couldn't be that bad? Wilf was 53, I was 52. Surely we had many more years together?

A couple of weeks later, after Wilf had already gone to see the doctor, I went out to get a paper. A woman in the shop asked after Wilf and I told her about the lump. It turned out her husband had worked at exactly the same place as Wilf in Ellesmere Port. Like Wilf, he was a fitter, putting lead into petrol; using Tetraethyl lead, known as TEL. The woman's husband had died. From lead poisoning.

I went home and told my brother. I wanted to see the doctor on my own. It was true. The doctor told me Wilf had a few months to live. That week, he was taken into hospital. He went in on the 3rd October and died on the 6th December. There were three men in our road that all died the same way, from lead poisoning. All working at the same place, though Wilf had left the firm before he got ill.

It took me years to get over it. You didn't know what happened in factories then. He got a medical every month. And around that time they stopped putting the lead into the petrol. But it was too late for Wilf. He was in terrible pain, my lovely husband, crying in pain. And we both pretended we didn't know. Because that was what I'd promised. After that I moved, to where I live now – Wilf's pension paid for the house. And I worked for 20 years as a supervisor in a laundry business.

For years after Wilf had gone I'd just go to work and come home. But one day I decided to go on an Alpha course, which looks at the basics of the Christian faith. And at the course, I realised I could share my experiences with other people. And only then was I able to talk about what had happened. Then I got involved with another group, a

Methodist group where you share your experiences, have a meal. That changed everything for me.

I'd always kept up my Guiding, always took pack holidays. So when I got a letter one day last year, to say I'd been awarded an MBE, it was a truly wonderful moment. I don't feel bitter about the war, about what happened to me afterwards. But I do feel cross at times that those of us that worked in munitions haven't had any recognition yet. Everyone else has had some kind of thank you.

Certainly, everyone around you was doing their bit for the war. If you went to the shopping centre in the middle of Birkenhead during the day, you'd only ever see children and old men. There were no young men at all; they were all away at war. I often used to think about that.

We had to make our way to the factory through the bombing, feed ourselves on rations – which was very hard – and we were in exactly the same danger, working through the air raids, as if we'd been in the Forces. No air raid shelter for us if we were on a shift. We had a man with a whistle up on the roof. They told us he was up there to reassure us but I think he was up there to blow the whistle if it was a gas attack.

You never got told anything in those days.

CHAPTER 6

LAURA'S STORY: AN ANGEL AND A ROSE

'WE WERE TINY COGS IN A SECRET ARMY'

Laura Hardwick was born in Trimdon, County Durham, in 1921. She has the double distinction of being both an 'Aycliffe Angel' and a 'Swynnerton Rose'. She spent two years making bullets at Aycliffe until she was called up and had to leave home to work at Swynnerton in Staffordshire, where she helped make detonators for the Navy. After the war she married and remained in the Bishop Auckland area. Her husband, Bill, died in 1996, age 80. She has one son, three grandchildren and three great grandchildren. This is her story:

The Germans knew we were there. But they never found us. Lord Haw-Haw [the nickname of Germany's propaganda broadcasters] was always making broadcasts, saying we had been bombed – he called us 'the angels in white coats' – but the truth was, the factories in Aycliffe could not be seen very easily from above. They built air raid shelters for us workers – but we were never in them. We just had to carry on working.

My very earliest memory is the day I started school, when we lived at Etherley Moor, just outside Bishop. One of the families living near us had a little girl the same age as me. So we started together, hand in hand, walking through the gate at Cockton Hill Infants School, not far from where I live now. I made friends with that little girl, of course. And then she died. When that happened, I didn't want to go back to Cockton. I was so upset, my mum, Emily and dad, Joseph, sent me to another school, Escomb. All I can remember of that is one freezing-cold, snowy winter when the older kids pulled us little ones to school on a sledge.

We were just an ordinary working-class family. Dad worked in the steelworks but when WW1 broke out, he and his brother were found to have hammer toes – which we all have – so he couldn't march properly. So he went to work in the pits for a while. I was the eldest. Ethel came over four years later and my brother Robert was nine years younger than me.

In those days, everyone you knew was poor. But we managed. It's not like you read in books, where kids had no shoes and were running around dressed in rags. We weren't like that. I was not brilliant at school. Ethel was the brainy one, she wound up going into nursing. I do remember one thing, as kids we always played shops for some reason. Ethel seemed too babyish to me, with nearly five years between us, yet when Rob arrived, I was nine years old and I mothered him. He always seemed to be with me as we grew up.

In those days, you left school when you were 14. There was

an 11-plus exam but I didn't pass it. There wasn't very much work around here, so I wound up working on a farm with a friend who was the farmer's niece. We bottled the milk for the schools in the area and generally helped out around the farm and in the farmer's house sometimes. The pay was five shillings a week. After a while, I found a job working in a grocer's shop; also five shillings a week.

I was 18 when war started and I'd been working at the shop for quite a while when it all happened. Where we lived was only a little village, really, so everyone was worried about it all in the beginning, buying the blackout curtaining, that sort of thing. Then came the food rationing. The idea of going into the Forces didn't appeal to me one bit. I'd never left home before, though of course in the end I had no choice; I had to. But the general idea was, everyone had to do something to help the war effort.

When they were building the factory at Aycliffe in 1940, my father left the pits and started working there as a labourer. Then, when it opened the following year, he went to work inside the factory. By then, nearly all the girls I knew were going to work at Aycliffe. The pay was £2 a week, which was a huge difference from my earnings. So when I was called up, I chose Aycliffe.

When I told the lady I worked for, she said she could get an exemption for me. She even offered to put my wages up to seven shillings and sixpence if I'd stay. I don't think she could have got an exemption, to be honest, though I think, as a farmer's wife, she thought she could try. But I'd made my decision.

My friend Dorothy was already working there. 'Ask if you can be put on Group One,' she told me, because that's where she was working. But when I told my dad this, he put his foot down. 'You are not going on Group One, Laura. Ask to go on 7A.' I didn't know what any of this meant. When I went to the Labour Exchange for my interview for Aycliffe, I asked the lady if I could go on Group 7A.

'Why?' she wanted to know.

'Well… me dad works there,' I explained.

That was it; she arranged for me to go to Group 7A. A long time later, I found out that quite a lot of girls working on Group One were killed in accidents. You hadn't a clue, really. Everything was just kept quiet.

On 7A we made rifle bullets. We worked at one end of the shop floor where there was a huge machine operating a big belt. The machine would slide the bullets down and our job was to press the detonator in as the bullets came along. From there, the belt gradually worked its way down the shop floor and eventually produced the finished bullets. We had to turn out thousands of bullets a day on every shift.

When I first started working there, we'd travel to work by bus. But then they built a special railway station near the factory. After that, a bus would collect us from the villages, take us to Bishop Auckland and then we got a train to Aycliffe. Then we would board the utility bus – 'Tilly buses' we called them – with wooden seats, waiting to take us to different sections of the complex where they'd drop us off at the gate; you'd show your pass and walk through to wherever you were working.

We worked in white turbans and overalls – no grips, no jewellery, no watches or rings, nothing. When you got there you went straight into the changing rooms, left your outdoor clothes and shoes on one side of the barrier, then once you'd crossed the barrier you got your shoes, overalls and turbans on.

I can still picture it now in my mind's eye. It was a very long bench with all these women sitting beside it, knocking the detonators into the bullets. I was working with girls and women of all ages. There were older ladies there too; they always fell asleep on the night shift, poor dears. Every so often, one of us would make a mess and then the belt would come off. So you'd have to find a fitter to fit the belt back on. You couldn't just press a button to do that in those days.

Oh, it was so monotonous, and it was incredibly noisy on the shop floor – all of us sliding the bullets down, the same thing all the time. To help break the monotony we'd start singing. That was all you could do to keep your spirits up. When we went for our break, we'd have *Worker's Playtime*, and people like Betty Driver out of *Coronation Street* would come to the factory to entertain us, sing and dance for us girls. Or you'd get people who'd make you laugh, like little Arthur Askey.

They were always checking us. We'd have men and women walking in, dressed in black coats. They would say, 'You, you and you', point to you and then they'd search your hair, check to make sure you weren't wearing any jewellery. They'd appear out of nowhere. It was, of course, a random security check. They had to be diligent all the time. And there were the other girls – the 'Blue Bands' we called them – who would walk around the shop floor, checking everything. We called the other sections of the factory 'the shops'. They were massive. They were so vast, you wouldn't have a clue what went on inside.

If there was an advantage, it was earning more money than I'd ever known. £2 a week was riches to me. I gave it to my mother and she'd give me back a little bit as pocket money. We knew that the lads who were fighting the war needed those bullets we were making: we had practically nothing in ammunition terms when the war started. I spent two years working in bullets.

There's one very vivid memory of Aycliffe that stands out. I often think about it, even now. Once, I was sitting on the train going to work. There was a girl next to me, moaning to everyone: 'Oh, I dunno, I just don't feel like going to work today.' That, to me, sounded daft. Did any of us really feel like going? What was the point of going on about it?

She must have been having some sort of premonition, that girl. Quite a while afterwards we found out that there was a young lad wheeling bombs on a trolley in another section and one went off accidentally and the girl on the train caught the blast. She survived, but she was very badly injured. It couldn't have been a very big explosion because the whole place would have gone up if it had been. But it was enough to ruin her life.

Not very long after that we got more news, something that was going to affect a lot more lives – including mine. We were informed that some of us young single girls would be having to move – to work at another filling factory at a place I'd never heard of called Swynnerton in Staffordshire. They'd be sending us there on a train and we'd be living in a hostel near the factory. Of course, none of us wanted to go. My mother wasn't very pleased, as you can imagine. A lot of us girls had never been out of Bishop – you never went on holidays in those days. I'd been on a school trip to Redcar and that was it. At nearly 21, I'd never even had a boyfriend. I was very shy.

But there it was, it was war work and single girls had to go wherever they were sent. We all turned up at Darlington station for the special workers' train and they issued us with red, white and blue labels to tie to our case, to show which part of the new building we'd be living in. The train took us from Darlington to Manchester, a hundred miles or so, not very far nowadays but a distance none of us had ever travelled before. It all felt very strange.

Once we got to Manchester, the WVS were parked there outside the station in vans, waiting to give us all a cuppa and a piece of spice loaf. Then we all had to get on another bus to the other side of Manchester and then get on a train to Stoke. Then we boarded yet more buses to take us to where we'd be living, the hostels. There were four different hostels – Raleigh Hall, Drake Hall, Frobisher and Nelson Hall, all a distance apart, for security reasons, and my group was going to Raleigh.

There was a little cottage at the edge of the hostel where the manager and his wife lived. Inside Raleigh, you were assigned to different 'houses'. I was in Juniper House: two rows of rooms with bathrooms and toilets at the end. You had a wooden bed with a wooden base, two girls to one room. I thought we had the best hostel because we had wash basins in the room with hot and cold water – a luxury for us. The room had two windows, one bed down one side, the other down the opposite side. There was even a shelf for your family photos to remind you of home.

The munitions girls living in the hostel came from all over the country, so there was a great deal of homesickness. At first, I shared a room with a girl who told me she was getting married soon. Then she went home to her village, a very poor place called Whitton Park. And that's what happened, she was married . If you got married, you didn't have to come back. So I was on my own for quite a while in that room. I didn't like that one bit – I wanted company more than I wanted privacy.

We all knew so little. We didn't know there were all these other munitions factories around the country. Now, at Swynnerton, I was sent to work on something called Group One. And it was much more dangerous munitions work than I'd done before. We had to go to a training area before we started. We were told we'd be making detonators for the Navy, though we never actually saw the finished product.

The lady that trained us was a bit of a shock to the system: when we all saw her, we wanted to run home straight away! She'd been in an explosion and had an artificial arm. She had also lost one eye. Part of her face still had shrapnel in it. And she walked with a limp. She stood there, in front of all of us, telling us what our new job entailed.

You were given a small tub of mercury, in powder form. The job was to fill the detonator with the mercury. The tub was pushed through a hatch to you and then you took the tub over to a 'hopper', a tray of detonators. Then you had to sprinkle the powder over the top of the detonator; then it went back into the hatch. You worked behind a steel shield with a Perspex window while you did the work. You also had to wear a cotton wool mask to do the work. This turned out to be so uncomfortable, it made your mouth sore as you filled the detonators. But, of course, you had to wear it. There was also a bench where we would be given special tongs to clean the detonators: each girl took a turn to do that as well.

It was incredibly dangerous work, filling detonators and

cleaning them. We were being paid two shillings and sixpence a week extra to work there – but we knew all too well how dangerous it was. We already knew that girls who had worked in Group One had been killed at Aycliffe, though during my time in Swynnerton, I didn't hear of any girls who had been killed in our section.

One very scary day at work stands out in my mind, though we were never told exactly what had happened. We were sitting on our stools, working away, when there was an explosion, though we couldn't quite see or hear where it had come from. All the 'danger men' rushed in, and we were ushered out to the canteen, given a cup of tea and the entire area, the shop floor where we worked, was sealed off. The next thing we knew, we were all being sent back to the hostel. Yet by the time we turned up for our next shift, the shop floor was ready for us to start work again. It was as if it had never happened.

Some of the girls were quite daring in the way they worked. There were a lot of Irish girls working there and, for some reason, they'd take risks, or that's what you heard, because you never really knew for sure what the workers in the next shop did; you knew so little beyond your immediate area of work. One day, someone told me these Irish girls had put through more detonators than they should have done – and one went off. The explosion had blown the girls right onto the clearway. I didn't see it happen. I don't think they were hurt. But the story was, they were trying to do too much.

One consequence of working with the mercury powder was something called 'The Rash'. Your face and arms went red and itchy. They didn't give you anything for it, but once you saw the doctor, he'd examine your rash and then he'd clear you to go and work in a different group called 7C. In that group, you'd be making parts for bazookas [a portable rocket-propelled anti-tank weapon]. You'd be making little cartridges, pushing them into one particular part of the bazooka. It was about as monotonous as it could be, but it was far less dangerous.

You'd work in that section for a few weeks until the rash calmed down. Then they'd send you back to the detonators and the mercury powder. At one point, I'd gone back to Group One for about a day when the rash reappeared. So the doctor told them to send me back to the bazooka parts job for good. And that was the end of my time in Group One: no more rash.

Then I was given a different job. I sat at a table on my own with a big oil can with a big spout. Inside the can was shellac, which was used as a moulding compound or sealant. My job was to put a drop of the shellac onto a small cartridge. You had to be very careful. If you spilled the shellac over the side of the part, it would be a reject; you needed a very steady hand. I don't remember ever spilling it.

At one point, I'd made friends with a girl called Lily; she came from Perth in Scotland and, like me, she was very homesick. We decided to ask the manager at the hostel if we could share together, and she said she'd see what they could do. In the end, we did share a room and that made a huge difference for me. We became really good friends. In fact, Lily and I are still friends to this day. She lives up in Fife but we remember each other's birthday, that sort of thing.

They did lay on entertainment for us at the hostel. They organised dances and the soldiers used to come with a band and play. On our weekends off, Lily and I started going out together to Wolverhampton. It was all new for me, that kind of thing, walking around a city centre, arm-in-arm. I wasn't used to city life. We'd go round the shops, have something to eat, so different from the familiar things I knew at home. Somehow, as young girls, our instincts told us we had to enjoy what we could, when we could. You couldn't dwell on the fact that you were away from home, missing your family. And no one knew what the future held.

In the village at Eccleshall where Cold Meece, the Royal Ordnance Factory, was located, the villagers found it difficult to accept us. The local girls made it very plain they didn't like the Swynnerton girls being there one bit. They said we were taking jobs away from them. So we were never invited into anyone's home locally.

One Sunday, a small group of us Swynnerton munitions girls decided to go to the local church. Not one single person said a word to us. We did make an attempt to be pleasant. We even invited some of the local girls back to the hostel, just to show them where we lived. But we were never invited back. It was that bad. They just couldn't accept all 'the hostel girls', as they used to call us.

The food they gave us at the hostel wasn't up to much. I still remember the huge bins in the canteen. You'd have to scrape the remains of your dinner into them after you'd finished, so the farmers could collect them. Nothing ever got wasted in those days, not a scrap. Lily's mother up in Scotland had hens, so she used to send us eggs. She'd wrap them in newspaper and post them in a shoe box. She managed to send them off to us quite often – and not a single egg got broken. Lily and I would take them up to the serving hatch in the hostel and ask if they could be boiled for us. At home, things weren't too bad for my family either, foodwise; my mother had an allotment with hens and rabbits.

My mum worried about me all the time, being so far from home. It was bad enough when I'd been working at Aycliffe. At one stage, in 1941, the Germans bombed Hartlepool, near the coast, just over 20 miles from Aycliffe and she'd stood on the step outside our house, watching the skies, thinking: 'Are they going to hit Aycliffe? Will our Laura come home tonight?'

At holiday times, Christmas and Easter, I'd go back home. After I went off to Swynnerton my father was moved from 7A to another section at Aycliffe, operating the conveyor belts. It was good going home to see everyone but it was so hard when you had to go back. When you did go into the town back home, you could tell the girls who worked with the yellow powder at Aycliffe because their hair was yellow. Even the turban couldn't cover it up at the front. And my dad lost his hair. At first it turned yellow, then it was gone. The men, of course, didn't wear the turbans to cover their heads.

When you'd travel back to Swynnerton, the trains would always be packed, mostly with soldiers. We weren't in any kind of uniform. Munitions girls could not wear any uniform outside the shop floor, because they said there was a risk of contamination. So when we got to the station, the WVS wouldn't serve us tea. Cups of tea were only for those in uniform. In the end, we'd pal up with the soldiers, sitting on their kitbags. That way, they'd get us a cuppa from the WVS. The soldiers knew what we did in munitions and that we were part of it all. In a way, little things like that made you feel you were involved with some secret or underground force. But you didn't want to waste your time resenting it: you were doing something worthwhile and that was what really mattered.

On one visit home, I went with Mum and Dad to see some relations in Middlesbrough. And there I was introduced to a George Hardwick. He was in the Royal Artillery and briefly home on leave. I wasn't that keen at first. George was five years older than me and had been born in Australia. But before we left, he asked me if we could keep in touch, write to each other. And that was the start of me and George. He wound up being sent to Malta, working on the big guns.

What with having a boyfriend and with Lily as a good friend and roommate, I suppose that kept me going with the work at the factory. But it was still a struggle. I'd get very down at times, missing my family. I struggled, being away from my home.

In the last few months of the war, when we all knew it was nearly over, Lily left work and went back to Scotland. By then, quite a lot of people at the factory were leaving. Some of the Irish girls at the hostel were courting Americans – and many of them were making plans to marry them and going to live over there. Bill too had now left Malta and wound up being stationed at Catterick. So by then, we were able to see a bit more of each other.

I actually left Swynnerton just before the war finished. I went home on leave and I was very down, knowing that Lily wouldn't be there when I went back. My mother, who was working as a cook at a children's nursery at the time, was so worried about me, she confided in her boss.

'Get her to see a doctor about the depression,' was her boss's advice.

I went to see the doctor and told him a bit about how down and unhappy I was feeling.

'We need to sign you off from Swynnerton,' was all he said. He'd be sending them a letter.

And that was how it all ended, my time in munitions. It had all gone on for too long, what with me being away from home and everything. Then, when Lily went, it just made it all worse. There was no official farewell at Swynnerton. I went home on leave and it turned out I never went back. So many of us then were completely exhausted, the war had gone on for nearly six years. Though I have to say now, working in munitions, dangerous as it was, didn't leave me with any major health problems. Ok, I had the rash, but nothing else.

We were so innocent then. You were still innocent when you got married, you just didn't have a clue. It's true that some girls would go with the Americans just to get nylon stockings. And our lads definitely didn't like the Americans, but that was pure envy – they had lovely uniforms. And more money. Our lads just had the rough khaki.

Bill and I were married at Escombe Church in May 1945. He was 28 and I was 24. By then I'd got a job working at the children's nursery where my mum worked, just helping out. Ten months after our wedding, I had my son Peter in 1946. That was difficult for Bill because he'd never been used to having children around him. He'd joined the Army at age 18 and he didn't have a clue – he thought a baby just lay there and slept all day!

We lived with my mother for a while, and then we got a prefab in Bishop Auckland, where we lived for a few years, then we moved to a house just round the corner from where I live now. We lived there for 60 years. We could have bought it but we preferred to rent it. I didn't go back to work until Peter was older and then I worked as a home help until I was 60. Bill and I retired together.

After the war, Bill didn't find it easy to get work; mostly he'd find work as an odd-job man. Then they started to build different kinds of factories at Aycliffe and he went to work there as a weaver. He wound up working at the factory at Shildon for 20 years, looking after 16 looms, making nylon, until he was made redundant. After that, he worked for nine years at Patons & Baldwins, the wool factory in Darlington, until he was made redundant again.

'I'm 60, I'll never get another job,' he told me. But he did. He worked at a local wallpaper factory until he was 65, and he was 80 when he died. He'd been in a home for nearly a year.

The funny thing was, we never talked much about the war afterwards, Bill and I. He always used to say that when a bomb hit a church in Malta – a Catholic country – the altar wall was always left standing. It was very, very strange. And when a priest went round visiting people, they all kept a hook outside their door for him to hang his umbrella, a sign that other people couldn't visit because the priest was there. Just little things like that, he'd tell us. But that was all.

My grandson's in the Marines and he's been in Iraq and Afghanistan but he doesn't talk about it. Some people don't. One relative was in the Falklands War and he discussed it with counsellors afterwards, because he got post-traumatic stress. 'I shouldn't be here,' he told me. 'I should've got that bullet.' This was because the lad behind him was killed. When times are really tough, I think sometimes people find it easier to keep it all to themselves.

There were parts of the war that weren't so bad. Those

weekends off, when we were at Swynnerton, were good, going to different places. And there was the entertainment they put on for us girls – the entertainers going round all the hostels, putting shows on. All kinds of things – singers, even ballet dancers. They did their best to keep morale up. There was even a tuck shop at the hostel, where you could buy cups of tea as you watched the entertainment.

We do feel a bit left out, us munitions girls. The Land Army, the Timber Girls, the Bevin Boys – they all got recognition. Perhaps it's because they all wore uniform, so everyone knew who they were. But we were Britain's hidden army. There was a lot of secrecy around what we did, but everyone in the area knew the girls who worked at Aycliffe. So it wasn't that secret, was it?

Looking back, knowing what I know now, it would be very difficult to go through an experience like that the way we did. Perhaps because so many of us were young and fairly innocent, that helped. That way we could handle the long shifts, the secrecy, the worry about the war, your family and the boys overseas. And if you weren't married, well, you had to do some kind of work.

There's one very clear picture in my mind's eye. I can still see us all now, getting off the buses that took us to work, going through the main gate, singing 'Bless 'em All' at the tops of our young voices as we made our way to our places on the noisy shop floor. We were just tiny cogs, the girls who made the thingummybobs, as the Gracie Fields' song put it. But at the same time, we had each other and we had our youth. So of course there were some good memories of it all, which we still look back on, even now.

We all knew you just had to make the best of it, you see.

CHAPTER 7

MARGARET'S STORY: THE TEASING GIRL

'WE WORKED SO HARD, WE'D WORK IN OUR SLEEP'

Margaret Proudlock was born in 1923 in rural Dalskairth, near Dumfries in the Scottish Borders. At 14, she went into service and by 1941, despite her dreams of joining the ATS, she started work in the cotton teasing section at the ICI Drungans munitions plant in Cargenbridge, where she worked for three years. Margaret's husband of 30 years, Roland, died in 1976. She has five children, 14 grandchildren and 11 great grandchildren. This is her story:

My grandparents were in service most of their lives. My grandfather, John Hutchinson, was Head Gamekeeper for Coats, the thread company. They had a lot of land. And ten big dogs.

I was born at the Gamekeeper's House at Dalskairth on the Dalbeattie Road, at the foot of a long wood. I was the second daughter; my older sister Donna was born two years before. Two other sisters, Jean and Alison, came after me. My father, Charles Welsh, delivered milk. He'd get the milk from his sister Agnes's farm at Lochfoot and then he would go round selling it off the back of a van.

He had a big metal container with two handles and a spigot to run the milk out. We had the only vehicle in the village. My father was a big, handsome man but he could be a bit naughty, going drinking with another milkman on the other side of town and then driving home the worse for wear. He had a habit of buying animals from the market just before it was finishing for the day. Once he came home with a van full of hens. He told my mother, Alice, he felt sorry for them.

We lived in a village house with no running water, a loo up the top of the garden and coal fires with a great big chrome fireguard. My mother would grow cresses and trumpet shaped flowers. One day, a bee got into my little puff sleeve and my dad ran out, ripped it open and the bee flew out. It's the little things you remember, your dad to the rescue.

We had another child living with us as time went on. My older sister Donna had to get married at 18, just before war broke out. Her husband wound up being captured at Dunkirk and was a PoW for five years. So later on, little June, their daughter, came to live with us all. Our school in Lochfoot village was just three houses up from our house. The headmistress was divorced: the village people had never heard of divorce then and they didn't take well to her. Yet she was a good headmistress.

Many of my childhood memories were of playing in the

Franky Wood, beside the Lochfoot Loch. We'd take old cooking pots down to the loch to wash, and we'd pick raspberries. I'd run home, get some sugar and we'd make raspberry jam, storing it in old meat paste jars. At other times I'd wander off and they'd be looking for me – only to find me fishing in the loch with a homemade fishing rod. I remember catching perch. The men used to catch big pike, a very rough fish. I still remember watching one man cut the pike open and seeing a dead moorhen inside.

My working life started in 1937 – in service. I'd passed my bursary exam that allowed me to go to the High School in Dumfries, but I left at 14 because my mother had found me a job. In fact, I often dreamed of being a nurse. But you had to do what your mother told you.

I started in service the day after I left school at Newtonairds House, a big house belonging to Douglas Menzies, a sheep farmer in Australia. The cook, who was ready to retire, trained me in lots of dishes. As the new scullery maid, I'd sit on a high stool beside her and learn. My first job, something I'd never done before, was to pluck 10 pigeons. 'Don't tear the breasts' I was warned. 'They'll have to be pot roasted to go upstairs whole'. Needless to say, feathers were everywhere.

The dinner service they used in the big house was trimmed with gold. So I had to be very careful when I washed it and put away. As well as learning how to prepare the family's food, I'd have to take the dinner trays along to the lift at the precise time and collect the empty ones when I took the sweet up. I'd have to look after the servants' hall too, set and make their porridge in the morning and their meals during the day. Sometimes you'd help yourself to the leftovers.

I worked there for over a year; then I went to another big house, Barjarg Tower, a very old house that went back to the 16th century. The Tower was about 14 miles away, so I had to cycle there and back. I was both scullery and kitchen maid, which kept me busy all the time; I also had to scrub the whitewood chairs and tables in the servants' hall every week. And I was still making the servants their meals and serving them.

When war broke out in 1939, evacuees from Glasgow were sent to stay in the house. It was all very confusing for everyone. The staff were moved from the servants' rooms to the gentry's changing rooms to make way for the evacuated people. In the end, though, the Glasgow people went home. They hated it in the big house out in what to them was the middle of nowhere.

One day, I went into my room and found that two older housemaids had written 'slut' in dust on my mirror. I cycled home that evening and told my mother I wanted to leave. Not long after, I was employed at Comlongan Castle, near Gretna, working for the Earl and Countess of Mansfield. They'd just got a new kitchen maid who knew nothing. Within a few weeks of starting there, all the staff, including me, had to go down to London to work in their town house in Cadogan Square. I still remember my high scullery window. I'd be working away, looking at all the people's feet going by on the pavement.

The war was starting to have an effect on everyone's lives. One day, the butcher's boy who came to deliver told me they were building air raid shelters everywhere. He had one at the bottom of his garden, he said. It felt so strange being so far from home. London was so big and scary, let alone thinking about the war. One night, the kitchen maid, who came from Carlisle, told me she'd had enough. She was leaving. Did I want to go with her?

That was enough for me. With our little cases packed, we slipped out of the door the next night. We went to stay the night with her sister who lived in London. The next day we were on a bus to Carlisle, and eventually I got home to Lochfoot. My mother was disappointed but pleased to see me. She'd had a telegram from the cook saying I'd run off. But I was really happy to be home.

My dad was too old to be called up. An old woman in the village got me a job in a hosiery factory. At 17, I was the youngest one there, making woollen gloves for the Forces. You made khaki for the Army, blue for the Airforce and navy gloves for the Navy. It was quite intricate work: a machine knitted the base of the glove and then you had to pick on so many stitches for the thumb and work out how many stitches for each finger. I got quite good at it.

But then, out of nowhere, something bad happened. At work, I started having these terrible stomach pains. The forewoman at work sent me off to the doctor. It was appendicitis. And unfortunately it had burst, so I needed a life-saving operation: peritonitis had set in. After the operation, I had a month in a convalescent hospital. There were soldiers there who had suffered snipers' bullets when they were on assault courses.

Once I got home, war work was on my mind: no one was talking about anything else now. It was 1941 and I'd turned 18. I'd seen the ATS convoys in Dumfries and I'd watched the girls in their uniforms and thought, 'Oh, I'd love to be one of them.' Yet when I told my dad about this, he wouldn't hear of it.

'Nonsense,' he said. 'Get yourself a job at ICI, like your sister Donna.' Donna was already working at ICI Drungans munitions plant near Cargenbridge, helping make gun cotton. I was still quite fragile from the appendix operation, but I went to the labour exchange anyway, and it was decided that they would give me a 'light' job in the cotton teasing department at Drungans.

That was my first job in munitions, while I got up my strength. I did it for about three months. We worked with big bales of cotton, feeding the wads of cotton into a machine that tore it up, a mechanical process. When it was close enough to fluff, we had to pack it all into three-foot-square metal boxes. Then we'd have to place the boxes onto trollies to be collected by the girls who worked in the nitrates section.

It was lovely and warm in the teasing department. Even when we'd finished our shift, we wouldn't rush out; we'd lie down on the floor, enjoying the warmth. And sometimes we'd all start singing 'You Are My Sunshine'.

I soon got used to the journey to work. We were picked up at 6am by a small utility bus, with slatted wooden seats, to get us there for the 7am to 3pm shift. By this time, I'd already decided I wanted to work in nitrates. You got double wages – which meant I'd earn about four times as much as I'd got in service. That, to me, was fantastic riches. I'd see the nitrates girls coming through in their wellingtons and rubber pinnies, coarse trousers and a bonnet with a wide band to tuck their hair in. It wasn't as comfy as the cotton teasing uniform, the trousers and woollen jumper with drill shoes we had to change into when we got to the factory each day but of course nitrates was completely different – it was dangerous work. I knew that. But I still wanted to work there.

What we had to do in nitrates was soak the cotton in big pans of nitric acid. You had to move the cotton around with a sort of prong to make sure it was completely saturated. Then you had to put stone plates, three inches deep, on top of the punched cotton in order to press the acid out. Then water would be run on top and the acid would drain away.

We worked on a production line. You'd press pan after pan, making the cotton brittle enough to turn it into gunpowder. This was heavy work, so we had to work in pairs. I'd put an armful of cotton in to soak and my neighbour would punch it. Sometimes, cotton with acid on it would come up and splash your face, so you'd get acid burns. And the acid would get onto your clothes, burn your bra, which would wind up in holes.

If it did burn your face, you'd have to run to the medical room and the nurse would put acriflavine [an antiseptic powder] on the burn for you. Then you'd be sent straight back to the huge shed area where we all worked. It was an enormous area, probably big enough to house 24 double-decker buses.

They kept a big barrel of water in the shed. This was because of the acid, in case someone got badly burned. If that happened, they had to throw themselves into it, to save their skin. But usually, the acid just splashed you in the face and you'd have to run like billy-o to get the stuff onto your face to calm it down. The ends of our fingers were always covered in little potholes where the acid had got to them.

I made some really good friends at Drungans. We had fantastic camaraderie. One friend, Sadie, her husband was in the RAF. She stayed with her parents and little girl for the duration of the war. The other ladies on our shift came from all over the area, wives and daughters mostly, from mining families: Ayrshire girls from the Nith Valley, Thornhill, Kelloholm and Kirkconnel. A lot of the girls from Dumfries were sent to work in the munitions in Dalbeattie, at the cordite factory. Sometimes, if we finished our shift early, we could all go and have a shower at work. We didn't have a bath in our house, so that shower was a luxury to most of us.

In the canteen, we'd all sit there listening to Churchill on the radio, saying, 'Today we lost a ship with 200 sailors' and when it was that kind of bad news, we'd all feel dreadful. Thankfully, Sadie's husband came back from the war but there were people from our village that didn't return. You were always conscious of the war. We'd cheer up, of course, when we had the entertainment they laid on for us, like the concerts from ENSA in the canteen at dinner time. And we'd often sing and play the piano in the factory after we'd had our evening meal. We kept ourselves cheerful that way.

One night, it was shepherd's pie in the canteen and I put a forkful in my mouth and went, 'Oo, this is a funny taste.' But I still ate the lot. Then one young girl – I don't recall where she was from – said: 'Paraffin food, paraffin food.' And she was right. It turned out a farmer had sent in a bag of potatoes where paraffin had been spilt all over them. But there were no ill effects.

Sometimes you'd get home from your shift and your skin would itch really badly. You could hardly sleep. But you couldn't take a day off or lose your work; you just carried on as usual. It was hard, heavy work, pushing the big trolleys of pressed cotton through to the steam room, where it was dried and turned into powder. It was physical effort all the time. There were times when we worked so hard, we'd work in our sleep: my sister Donna and I would throw our arms around and hit each other in our sleep – we were still working!

When my shifts permitted, I'd go to dances in the village. My mum would play the piano and we'd have concert parties where everyone would sing and dance to the 'Trolley Bus' song ('clang clang clang went the trolley') or 'Lily Of The Lamplight' ('Lily Marlene'). I had a wartime boyfriend too for a short time: Tom Hicks, he was in the Royal Yorkshire Signals. Tom was a good dancer and a great skater when there was ice on the loch. But then he got sent abroad. We managed to write once or twice. Then he met a girl while he was abroad.

One day, I was at home, sitting at my mother's sewing machine, taking up the hem of a coat. I had flat metal curlers in my hair. Then my sister Jean came running in.

'Margaret, someone wants to meet you.'

Jean's boyfriend had turned up at our house with another boy.

So there I was, outside our home, still with the metal curlers in my hair, being introduced to Roland, age 19, same as me. A nice-looking fellow, he was a farm worker, so he hadn't been sent off in the Forces.

'Are you coming to the dance Friday night?' he asked me.

You bet. I cycled to the dance that Friday, me in my nice new dress and shoes, my holey bra, burns on my face covered with acriflavine (which turned the burn brown). But my war wounds didn't seem to matter a bit to Roland. Soon, we were seeing each other all the time. My parents were a bit worried about it all at first. I was only 19, after all. On our first date in the town they even watched us meeting up to go to the pictures, standing far away enough that we couldn't see them, my mother told me later. But he soon won them round.

The routine at work was one week night shift, one week day shift so going to the Saturday night dance depended on your shift. But no matter what, you had to be in by 11pm, my dad said. Donna's husband was a prisoner-of-war and my dad would get angry with her if she came home late.

As far as I knew, there were no serious accidents in the factory. But you also had to work with a constant threat from the fumes from the acid as well as the burns. If a tiny bit of water got into the cotton treated with acid, the pan of acid and cotton would catch fire. Thick yellow smoke would fill the shed and go up through the slatted roof. When that happened, we all had to rush outside. And if it was your pan that caught fire, you'd have to lose your bonus payment. The security was extra tight, too: you'd be searched coming in and going out to see if you were carrying cigarettes or matches. But our attitude to it all was, well, we've got a job to do. We were helping make the ammo for our boys to fight in the war. And the thing was, even though the job itself could be dangerous, we didn't feel it was as dangerous as fighting at the front. I was really proud that I was doing my bit.

I was 21 when the war ended, after three years working at Cargenbridge. At first, we couldn't quite take it all in that we had actually won. I'd already been engaged to Roland for quite some time. About a year after we'd started going out, he'd turned up one night with a turquoise ring – he'd already asked me my favourite colour. I was thrilled to bits.

I took the ring to work and gave it to the super to look after while I was working. Roland said he wanted us to marry after the war. So now, the news that the Germans had capitulated meant I could start planning our wedding.

For my sister Donna, it meant her husband Graham was coming home. We put out the welcome sign for him but he looked so thin when he returned after all those years in the PoW camp. He'd had bronchial trouble beforehand, so being a prisoner for all those years didn't help his health.

ICI closed down as soon as they announced war had ended. No more clocking in with a policeman standing by for every shift. It all ended without any fanfare at all. One day we just clocked out – for the last time. It was over.

We got married in August 1945. It was the first wedding in Lochfoot village after the war. I went round all the farms in the area and bought extra clothing coupons, and a woman who lived two doors down made the bridesmaids' dresses. I managed to borrow a lovely oyster satin dress. The whole village turned out. In a way, it was a double celebration, our marriage and no more war. For my going away, I wore a mustard-coloured linen suit with a three-quarter jacket. I was crying – I was leaving my family, my parents, everyone I'd been so close to during the war. It was the end of something. And we didn't go for a honeymoon, as such. Roland had a wee racing car, a Midget, a two-seater, so we drove off in the 'selfish wee car' to a hotel in Newcastle.

Within a year, our daughter Louise was born. Two years later, Alison arrived. Roland managed to get a farm cottage until we settled in the farmhouse, three miles up the road from Shawhead. Louise was just a baby when we got the cottage half a mile from the farm: outside loo, no running water, you'd go up into a stream in a field for water to drink. Roland worked hard on the farm. He also played percussion in a dance band and he was a great singer. So some weekends I'd be left alone. Then in 1951 we moved to the house in Shawhead, where I still live now.

Elizabeth arrived 18 months after Alison and then Roland junior came three years after Elizabeth in 1953. My youngest, Malcolm, arrived in 1960. By this time, Roland had taken on an agricultural contractors' business – he got all the machines, the combine harvesters, the tractors. He'd buy the machines and the local farmers and farm workers hired the machines from him so that they could farm. It got so busy he had to have three men working for him.

None of my sisters learned to drive, but I did. Sometimes I'd have to drive the men out to where the tractors had been left in the fields. The farmers would often ring us at 5am when they needed a machine to get started. So it was always a hectic life, helping Roland, looking after the children. If it was dry, Roland could often work till 11pm.

In 1971, we went on a big holiday to Australia. Roland's brother, David, had gone there for £10 on an assisted passage so it was high time we visited him. Our bank manager was shocked when we said we were going to Sydney for two months. 'How can you afford that?' he said. We could. The business was going very well at the time.

It was just as well we made that big trip, because five years later, in 1976, one frosty morning, everything was set to change forever. Something was very wrong with Roland. The men who worked with him were so worried, they came into the house and spoke to me about it.

'He keeps sitting there, holding his head, Margaret, but he

won't tell us what's wrong,' one of them said. Then, when I started pushing him to tell me, I got the truth: Roland admitted he was having the most terrible headaches.

'But I'm not going to any doctor,' he warned me.

'They only tell you there's something wrong with you,' were his words.

The next night he'd been out playing at a staff dance in Dumfries and didn't get home until 3am. Even then, he insisted on going out to collect some coal. Then, as he bent over our coal bin, he clutched the back of his head suddenly. I could see he was in terrible pain.

'Margaret... I can't see now,' he told me. Then he collapsed.

I rang for the doctor and after he'd looked at Roland lying there, semi-conscious, he took me into the kitchen.

'It's a haemorrhage, a burst artery,' he said. 'There's nothing I can do for him.'

I went with Roland in the ambulance but when we got to the hospital they made me wait while they took him up to the ward. When they finally directed me to the ward, I found him unconscious in the bed. He remained that way for three days until he died. All the family came and rallied round us, but there was nothing anyone in the world could do. We'd been married for 30 years.

It was tough, afterwards. We had to close the business down. One of his brothers was a partner but he couldn't drive. Roland always did the books. But I knew I just couldn't run the business on my own. With the help of my daughter, I managed to collect all the money we were owed by the farmers. Then we managed to sell a lot of the equipment. Roland had bought three new tractors but they weren't yet paid for, so they had to go back. There was a lot of money lost, too.

It was very strange afterwards, but my three daughters were all expecting at the time Roland died, so the next year three babies arrived. That made a huge difference. Now, I can say my children have all done well in life. And I have 14 grandchildren and 11 great grandchildren.

I don't mind living alone at all. Elizabeth is four miles away and Malcolm is the same distance. I'm out a lot, anyway. And healthwise, there aren't many problems; maybe a pain in my hip when I climb the stairs, but that's it. I still do the Scottish country dancing every Monday night. And I drive myself to church every Sunday – I never miss that.

To be honest, I don't think the world is now any better than it was all those years ago. My children were able to play out like I used to when I was young. Any mother in our area wanted to know where her child was, they were here. That's exactly how it was. But nowadays people won't let their children out to play and they've all got their computer machines. No conversation, all playing with these machines.

There's such a difference, living in the countryside. To me, even Dumfries is like living in the big city. Some folk smile at you in the street, but in the main, you're not part of a small community, the way you are round here.

I'm proud of what I did in the war. I'd describe myself as someone who tries to do everything as well as they can do it, and who tries to do as much good as possible. My only regret in life is losing my husband the way I did. But then, when I think about it, I wouldn't have liked him to be an invalid for the rest of his life.

My Roland would have hated that.

CHAPTER 8

MAISIE'S STORY: MAISIE FROM ESSEX WITH THE FACEPACK

'ON NIGHT SHIFT, I'D GO INTO THE TOILETS AND FALL ASLEEP'

Maisie Jagger was born in 1922 in Woolwich, Southeast London, and grew up in Dagenham, Essex. She worked as a shop assistant and a machinist until war broke out. In 1940, she was conscripted for munitions work making gun cartridge cases at the small arms ammunition factory in Blackpole, Worcester. Like thousands of other women, leaving home and relocating to a strange environment proved to be a very difficult, unsettling experience for Maisie. After 18 months at Blackpole, her health deteriorated and she was moved back home. Towards the end of the war she continued her munitions work in Dagenham, making parachutes and masts for dinghies until she married her wartime sweetheart, George, in 1945. George died in 2007. Maisie has one son, three grandchildren and four great grandchildren. This is her story:

People often ask me, 'What was life like then, how did you manage to live with all the rationing, the bombings, being away from your family?' I always tell them the same thing: you just got on with it; you accepted it; you couldn't do anything else.

I was one of five: two boys and three girls. My parents were Londoners, my mother, May, was from Leytonstone in East London. My father, Albert, worked as a labourer at the *Daily Mirror* newspaper in Holborn. At first we lived in Woolwich, Southeast London, but Dad managed to buy a house on a new estate in Dagenham, Essex, so we moved not long after I was born.

Our youngest, Jean, arrived 14 years after me. Jean was always with me, all the time, wherever I went, when she was little. By the time I was a teenager, if I went to a dance or to the pictures, I went with a crowd – but little Jean was always with us. We were a very close-knit family, that's for sure.

At the top of our road in Dagenham there were quite a few shops. I left school at 14 and went straight into shop work. I was never once out of work. I think I worked in every shop near us: a grocer's shop first, then a shop called Perk Stores, then a shop called Gunners – they had biscuits in tins all along the front of the shop. I worked on the counter of a fish and chip shop and I also worked in a place called Maypole. I can remember the cheese they sold, covered with something like a sack, or a type of webbing all around it.

I was quite a friendly girl. You had to get on with everyone in the shop, didn't you? But I'd also change jobs quickly for a penny or tuppence more a week. By the time war broke out, I was working as a machinist, making haversacks and binocular cases in a big factory in Oxo Lane, Dagenham.

Whatever I earned went straight to Mum on a Thursday – when I knew she wouldn't have any money left until Dad got paid on Friday night. In return, I'd get sixpence a week as pocket money.

Every Saturday night I'd go dancing at the British Legion Hall in Dagenham: the foxtrot, the waltz – I liked dancing, all right. I won't say I was good at it but I was dead keen on it. One of the daftest things I always remember is if I was going dancing and it was raining, oh how I hated it because all the mud would splash up the backs of my legs. I'd have to rush into the cloakroom, pull down my stockings and wash my legs, pull up the stockings and then out to the dance floor… it's quite daft the things you remember. But you do.

My mum worked in the police station as a cook until the war started. Then she had to be evacuated to Ilfracombe in Devon with my sister Queenie, my brother George and little Jean. My oldest brother, Harry – we used to call him 'Chink' – was two years older than me so he was old enough to be called up. Almost overnight, this close family, the Rushbrooks, was scattered, leaving just me and Dad in the house, which I hated because I missed everyone so much. Dad was very lucky – or unlucky, depending on how you saw it. He'd just finished 12 years in the reserves for the Navy when war broke out so they didn't want him.

'Chink' was really unlucky. He was captured almost at once, in Africa, by the Italians. He hated the Italian PoW camp, then the Germans took over and he wound up as a PoW in Germany. We got letters from him sometimes but it was all very unsettling for everyone.

I had wanted to go into the WRNS ('the Wrens') but my mum put her foot down, she wouldn't hear of it. Mums and dads used to be able to say what they wanted you to do and you did it. They had some funny ideas then. A lot of people believed the Forces were bad places for a young girl. 'The Navy's no place for a young woman,' my mum kept saying. I didn't know what she meant but there was no way I would disobey her.

In the end, of course, because I'd turned 18, I went to sign up for war work. At the Labour Exchange, with a group of other girls, we were told we'd be 'directed labour' doing war work. In other words, you'll be working wherever we want to send you. It turned out I was being sent to Worcester, nearly 150 miles away. My best friend, Lily, was in the same boat, also called up as directed labour. But she was being sent to Rugby, which was a bit of a blow, though we managed to keep in touch throughout the war – and have done ever since.

Naturally, I didn't like the idea of leaving home one little bit. I don't think anyone in that situation liked the idea of it, if you had a decent home. It was bad enough the boys being called up. We were all so close at home, especially me and Chink. Before war started, we had our own little arrangement when I went out at night, because our parents were quite strict with me. Dad wouldn't let me be out after 9pm, so Chink would say: 'It's ok, Dad, Maisie and I are going out together; we won't be back until 10.' Then Chink and I would stroll to the top of our road – and go our separate ways. We'd meet up again at 10 to walk back down the road together. I was always going dancing; I had no idea where my brother went.

When I think about it all now, leaving your family and going wherever you had to go was a bit of a shock to the system. We went to Worcester by train, with Dad waving me off at Paddington station; there were a lot of other young girls like me being waved off by their families. I remember the train was also crowded with soldiers; everywhere you went on a train in those years, it was always packed with soldiers with their kitbags in the aisles.

I sat myself down in a seat with my case on my lap and soon I got chatting to this very pretty girl sitting opposite me. Eileen Smart came from Rainham, also in Essex, so we had something in common. From that point on, Eileen and I stuck together like glue, two young women in a totally unfamiliar situation, no idea what lay ahead.

After the long train ride, there was a bus waiting for us and finally we arrived where we'd be living: a house in Lansdowne Road, Worcester, the home of a Mr and Mrs Prosser. He was a policeman. They'd arranged with the authorities to rent out their spare bedroom to two munitions workers: board and lodging, they called it. We'd be paying the Prossers out of our wages.

So there I was with Eileen, a stranger, in a strange house in a different part of the country. We'd be sharing a bed, Mrs Prosser said. I'd shared a bed at home with my sister; we had a double bed. But I didn't really like the idea of sleeping in the same bed with a total stranger. There was no point in saying anything, you just had to get used to it. As it turned out, me and Eileen were working on the same shift at work. The house itself was a three-bedroom house, quite sturdy. Mrs Prosser worked in a factory nearby and they had a son who would go away a lot and come back. The general idea was that I'd be able to go home to Essex once a month. But in the end, what with work and the rent taking quite a big chunk out of my wages, I only wound up going home about three times all the time I lived there.

The factory was in Blackpole, a train-ride away from where we were living. You had to wear an overall for the work and, of course, a hat, to keep your hair back. I was used to overalls from working as a shop assistant, but I hated the hat. The colour of your apron and your hat denoted whatever your shift was: blue, brown or green shift.

What we did was work with little bullets made of thick brass. They had a hole at the top. I worked with a machine that went round and round all the time in a big plate of soapy water, which ran down the side all the time. As the machine went round, you had to take the bullets and push them into the holes set into the plate. That was the work, day in, day out, pushing the bullets into the holes in the plate. The machines were big, like great big mechanical hammers. They would bash the thick brass into a long shape to create finished brass bullet cases.

But the actual filling of the bullets wasn't done where we worked. Many years later I found out that they filled the cases at the big Swynnerton munitions complex in Staffordshire: at the time, though, you wouldn't ask any questions about what it was you were doing. It was *so* noisy in that factory. You sat down to do the job and the girls all sat quite close together while we worked. But you couldn't hear one another properly if you talked. You couldn't hear the radio either, there was too much noise going on, what with the machines.

Eileen and I got on very well, probably because we were both similar; we desperately wanted to be back in familiar surroundings with our families. The other girls were ok, there were a lot of girls from the North there and we were told – and I don't know if this is true – that the bosses preferred the girls from the South because they were a bit sharper on the uptake about doing things, a bit more 'with it'. Well, the managers seemed to think that.

It was an all-female production line. The men came in to repair the machines if they broke down. And they'd clean the floors. But there weren't that many men working there – they'd all been called up.

I missed my family all the time. I hated the noisy factory and the night shifts, but I also had my boyfriend, George, to worry about. George was the same age as me. I'd met him in 1940, walking along Rainham Road. I was on my way to a dance and he just started chatting to me, a young uniformed soldier, taking his chances to chat up a girl.

George had joined the Territorial Army just before war was declared, then he was sent to Woolwich for training. Then he was posted near us, to Hornchurch, as a gunner in the Royal Artillery. What I didn't know that night we met was although he'd seemed very keen to accompany me to the dance, he wasn't in the least bit taken with dancing – secretly, he was hoping that this nice young girl he'd taken a fancy to couldn't dance. He only told me this many years later.

But it didn't matter. After that first dance we went out together for about six weeks until he was sent away up north. Then, we kept writing to each other. We wrote letters for about five years – I didn't see George again until after the war. After I started working at the Worcester factory I got a letter from him to say he was out in the desert, miles from anywhere, in the Middle East. So when I wrote back I mentioned my munitions job, but it seemed silly to go on about how much I hated it, because neither of us could do anything about it. There was no point in complaining to each other, was there? We were both doing what we had to do – just like everyone we knew.

It was a struggle for me, what with the shift work and the homesickness, those 18 months in Worcester. We did have entertainment every now and again, in the big canteen: live *Worker's Playtime* or a singer. But I found the routine of the work very tiring. When I was on night shift, I'd go into the toilets, sit under the sink in the toilet, no pillow, nothing – and I'd fall asleep for 15 minutes or so. I couldn't have slept for very long, how comfortable can it be, sleeping on a stone floor? And they weren't like the toilets we have now. They were made for workers in a huge factory; big, stark industrial toilets is one way to describe them. But that's an indication of how exhausted I was.

Food was a bit of a problem too. At Mrs. Prosser's house, you got something to take to work for lunch and a cooked evening meal, if you weren't on night shift. I can still remember eating Mrs Prosser's lunch in the big canteen at work: four tiny sandwiches, mostly cheese, she'd hand it to me in this OXO tin. I'd sit there in the canteen, almost as noisy as the factory floor, nibbling at Mrs Prosser's sandwiches. All too often I wouldn't even have the appetite to finish them.

I couldn't eat much of the food she gave us at home, either. The food in Mrs Prosser's house was nothing like I was used to. It wasn't her fault, really. There was a war on. People just had to make do with what there was at the time. My mum was a good cook, which made a huge difference, even if you couldn't buy everything you wanted. But at Mrs Prosser's, she never had much imagination with food, she was used to just cooking for the men, plonking it down for them and they'd just scoff it, without complaint.

At her dining table you got mostly potatoes, overcooked greens, plus a little bit of meat or something off the rations. Plus rice pudding, there was always rice pudding afterwards. I have never eaten it since. One time, I wrote a long letter to my mum, pouring my heart out to her about the food, and she even wrote back with a nice recipe for Mrs Prosser. I handed it to her and she took it. 'Oh thanks,' were her only words. But the end result was just the same: more boiled-up food. I suppose it was edible. Just.

The only girls in the factory that seemed to be really ok with it all were the ones who had come from a tough background: some of them made it quite obvious they actually liked being away from home, having their freedom, going where they liked on their days off. And, of course, having money to spend, which was a big deal for a lot of the girls. But I didn't feel that way: I wanted home. Yet even my visits home weren't very successful; the house seemed strange with everyone evacuated to Ilfracombe apart from Dad.

One day, I was called into the main office. The doctor at the factory had examined me a few days before and because I had lost a lot of weight and looked poorly, a decision had been made.

'You have to go home,' they told me.

I didn't put up any argument, that's for sure. They'd decided I just wasn't healthy enough to carry on doing the work I'd been doing. In a way, it was quite heavy work. But oh, how pleased I was to get on that train heading back home! I told myself I didn't mind where they sent me after that or what I did – as long as I'd still be living at home. Eileen was a bit upset that I was going, but as it turned out, she was fine. Another girl was sent to stay in the house and they got on well.

Once home, I was sent to work in Crolleys Factory in Oxlow Lane, just down the road from Sterlings, where they made machine guns. [The Sterling Armaments Company manufactured the famous Sterling sub machine gun used by the British Army for over 50 years.] At Crolleys, I worked on machines that were making parachutes and portable nylon inflatable dinghies, which air crews could strap on, if necessary. In my section we were making the thin, aluminium masts that were folded up inside the dinghies. I was still doing war work, working the same kind of rotating shifts I'd worked on at Worcester, but here I could walk to work. And I earned good money. I've still got one of my paypackets: £3 12 s 6d for the week.

My time at Crolleys was so much happier. A great crowd of girls there, always laughing and joking. Some of the cheekier girls would even put little notes inside the dinghies for the fighting men. They'd write silly things like: 'I'll be waiting when you come home', or 'My name's Elsie'. We didn't have phones at home then, otherwise I'm sure some of them would have put their phone numbers in too! I didn't do that sort of thing; I was writing to George all the time.

One Saturday, I'd just left work early because I was going to a wedding. George's dad, who lived in Wood Green, North London, had phoned through to the factory to tell me they'd had a letter. George was still in Italy, but he was ill and in hospital. The manager at Crolleys took the call

146

and sent a messenger boy out, trying to find me – they were really good to us girls working there. It was bad news: George had a burst ulcer. It had happened while he was on board a ship heading for the Second Front, but of course, he never got there. He'd been taken off the ship in Naples. As it turned out, he stayed in hospital until the end of the war.

In 1944, Dagenham was still very much in the firing line. We'd had a hard time during the Blitz in 1940 because there were so many factories in the area. Now London and its suburbs were under renewed attack from the lethal V1 flying bombs, known as the buzzbombs or doodlebugs, and a bit later, the V2 rockets. That was terrible. Some people who'd been evacuated from our area and had returned home actually packed up and left again, back to the country. More upheaval. It was a nightmare.

Dad used to get in before me of a night time. He'd be down in the air raid shelter in the garden. He'd have the tea ready for me – and we'd stay down there. Dad even put a hammock up in the shelter and we had what we called a 'bed chair' down there, an old wooden chair that could be made up into a bed. So I'd come home from Crolleys, go straight to the shelter and stay there all night. Then I'd go to work in the morning. That's how it was for millions of us. It was hard to handle sometimes at the end, what with the air raids and the noise all the time. It seemed like it was never going to end.

One night, in the shelter, I almost lost it. 'I can't bear this, Dad, I just can't bear it,' I told him. Then I started crying. The noise from the planes overhead, the bombs exploding, the horrendous clunking and rattling noises coming from the trains going down the railway line just behind our house were unbelievable that night. It was the worst I'd ever known it.

'Don't worry, Maisie,' my dad said. 'They're pulling the guns up and down the railway because they're going to use them to shoot at the planes.'

That calmed me down a bit. I have no idea if it was true; I think he made up the first thing that came into his head, just to calm me down. But there were lighter moments. I was a bit dozy sometimes. One day, Dad said to me: 'When you come home tomorrow, buy us some Edwards' Desiccated Soup.' That was a kind of soup cube people used then. But of course, I couldn't find it anywhere. So I bought a Foster Clark's cube and put it into the soup, telling him it was Edwards', which shows you how daft I was because Edwards' was dark brown when it was dissolved, whereas Foster Clark's cube was bright orange!

Of course I blithely insisted I'd done the right thing but it was the colour that gave me away. In the end, we had a row over it, the only time I ever remember arguing with my dad. How could he know? I wondered. But I still stood my ground and argued back.

'You try leaving work on a Saturday and trudging round the shops looking for Edwards,' I told him.

'I'm just saying, Maisie, it's not the right one,' he kept saying.

George used to write lovely letters and little poems while he was still away. Here's one of the poems he wrote me from hospital, written on the back of a postcard I sent him with my photo:

Good night and God bless you
May Angels caress you
And always watch over you for me.
Both waking and sleeping
My love's in your keeping
I pray in my arms you'll soon be.

In the end, George was taken home on an American cruise ship and I got a letter to say he'd arrived at Fazakerley Hospital in Liverpool. Then he was being transferred to a military hospital until he could come home. Years afterwards, he said he always kept a soft spot for the Liverpudlians, how the bus conductors wouldn't take their fares when the men got on the bus in their blue hospital uniforms [the light blue overalls signified that they were uniformed troops, still in hospital].

By the spring of 1945, the ending of the war seemed to happen quite suddenly. Everyone was coming home. The rest of the family got back from Ilfracombe, George was out of the hospital, and Chink had been released from the PoW camp: he'd made his own way back and walked hundreds of miles to get home.

Amazingly, Chink and George wound up walking down the road to our house together! By sheer coincidence, they'd actually come home on the same train from Liverpool. It was so lovely, my sister Queenie ran up the road to greet Chink – and there was George! We hadn't even had time to put out the bunting, it just happened.

George didn't waste his words. As soon as we'd hugged each other he said: 'We're getting married next Saturday, Maisie.'

'George,' I said. 'We can't do that!'

'We can. It's already sorted,' said my husband-to-be.

In the end, of course, it was me running round doing everything. You could get married by special licence without the banns being read so that's what we did at Dagenham Old Church on 4 June 1945 – less than a month after VE Day. There were so many of us, all across the country, who'd waited and dreamed of this day. There was a queue at the church: 14 couples had decided to tie the knot, now it was all over!

My son, David, was born in April 1946. We moved into rented rooms in Walthamstow at first, then once David arrived, we moved to a flat in Wood Green in North London. In the early fifties, we moved again, this time right into the heart of the City of London in Gracechurch Street. George had found a job working as a messenger for a City bank and wound up as a housekeeper for the bank. The job came with a flat and we lived there until George retired in the 1980s.

George, my wartime sweetheart, died in 2007. We lived for years in a little semi in Enfield but after he went, I wanted to move back to Essex, to Braintree which I'd remembered from years ago.

To be honest, working in Worcester wasn't bad in itself. The other girls were ok. All those years on, I think it was me. Just being away from home and everything I knew, I missed family, that's what it was. I knew we had to do the munitions work, but I loved my home so much. And I have never been a big eater, so what with not liking the food at the place where I lived and pining for my family, I hardly ate anything. No wonder I was so skinny: I was pining for home.

I wouldn't have been alone in that, of course. So many of us were separated from our families, waiting for our sweethearts to come home – you can only imagine what it must have been like for the men, so far away, not knowing if they'd get home, if at all.

A few years ago, my son David took me on a trip down memory lane to Worcester.

'Shall we go to where you were lodging?' he said. I pointed him in the direction of the street where the Prossers lived and he said, 'Oh, Mum, I bet you can't remember the house.' But I did. We knocked on the door and a man opened it.

'Hello,' I said. 'I was stationed here during the war.'

The man said: 'Well, a lot of girls were stationed here, but the only one I remember was called Maisie.' I couldn't believe it; nor could David. It was a bit upsetting, really. The man's mother, Mrs Prosser, had died just the week before.

He was around my age. He said to David: 'I've never forgotten your mother because your mother and her friend were always putting facepacks on. And one day, I said something daft and made your mother laugh, and the facepack cracked and she chased me round the street!' Oh, those facepacks! How we used to put them on and walk around with them. They set like concrete, so the last thing you ever wanted was a crack. To think we really did believe they'd make a difference.

I really don't know if other families were like ours in being so close. Some girls in my factory were glad to get away from home because young girls were very restricted then, in what they could or couldn't do. But for me, the separation meant that some of my war work was not a happy experience.

Nothing to do with the job, it was just me.

CHAPTER 9

ALICE'S STORY: FANCYPANTS

'EVEN THE WHITES OF YOUR EYES WOULD BE YELLOW'

Alice Butler was born in Burslem, Stoke on Trent, in 1925. One of a family of seven growing up in The Potteries, she started work in a tile factory at 14 – until she was called up to work at ROF Swynnerton, the big filling factory some 17 miles from her home. Alice filled bullets and smoke bombs at Swynnerton for over three years – until she was moved to a section filling explosive charges for the last nine months of the war. At 21, she married Tom Porter, a blacksmith in the local mine, Whitfield Colliery. Tom died in 2005, age 88. Alice has two daughters, two granddaughters, two grandsons and two great granddaughters. This is her story:

We were very poor as I was growing up. My youngest sister, June, always used to joke that if we were robbed, the burglar would have left *us* something – he'd have felt that sorry for us. There were seven children in our home – I was the third. My dad, Joseph, was a miner, but there wasn't work for him all the time until the war broke out and production was stepped up. Dad was also a rescue man for the pit.

This was how we lived: no electricity, only gas lamps; an outside toilet. Going out there in the middle of the night in the freezing cold, you'd try to hold yourself in if you could, it was that cold. A tiny little kitchen with a well in it – that was your water supply. A proper wash in a tin bath in front of the coal fire once a week.

Yet my dad was a fantastic cook: he tried to join up in the First World War, but they said he was too young, so they'd put him to work in the kitchens. As a result, Dad could make a tasty meal out of virtually anything. He didn't earn much at all but we never went hungry.

At Middleport School in Dale Hall, near Burslem, what was I good at? Well, history and domestic science. We were allowed to cook at school and that meant something to us because I could take what was left home to the family, so they could all eat it. That gives you an idea how poor we were.

If I tell you we all lived in a two-bedroomed house, two adults and seven children, you'd probably think 'how awful' – but you'd be wrong: we were happy. If you don't know otherwise, you get used to things like sharing a bed all the time. My sister Marina, who was 10 years younger than me, used to pee on me sometimes. I'd never slept in a bed on my own in my entire life until my husband Tom died.

Our neighbours were smashing. You never locked your door, there was no need. If anyone was in trouble, they didn't gossip about it behind their hands, they helped each other. My best friend, Freda, was in the same class as me and we became firm friends. And my three brothers, Joseph, James and George, always looked out for me. When I grew up and started going out dancing, James would always come and meet me afterwards. They were very protective.

At 14, I left school on the Friday and started working on the Monday at H & R Johnsons, a tile factory, at four shillings a week. By then, Joe and Jim were working in the pit, which meant money coming into the house regularly – for the first time, really. So by 1938, the year I was starting to work, I actually had enough money to buy new clothes; I could start to dress up a bit. I always loved dressing up – I still do. My nickname was 'Fancy Pants'!

All the talk around was of war but I don't recall we were too troubled by it – until it actually happened. The day war was declared, Dad, Joe, Jim, George and I were at Burslem Swimming Baths; we'd all learned to swim at school. We literally came running out of the water, dripping wet, at the point the news came through on the radio. Dad was very upset. I couldn't understand why at the time, but years later I discovered the truth: he'd lost his brother William at the Battle of the Somme in the First World War, when over a million people were killed.

For me, a 15-year-old, the idea that we were fighting a war was bewildering. And upsetting. Freda had gone to work at a place called Kents in Burslem, where they made elements for gas fires. By the time I was nearly 16, Freda was earning more money than me in the tile factory. So I decided to leave Johnsons and go to work for Kents. I was at home, between jobs, when the letter came saying I'd been called up. I was not yet 18 but the letter told me I would be employed in munitions at Swynnerton, the big filling factory. Had I not decided to switch and carried on working at Johnsons, who knows? I might not have been called up.

At first, when war broke out, I'd liked the idea of being a Land Girl: I've always loved gardening. The idea continued to appeal to me even once I'd started working at Swynnerton. In fact, even after I'd started there, I went down to the Labour Exchange a couple of times and asked if I could switch jobs and work as a Land Girl. But the answer was, 'No, your work's more important'. It was too late. But then, I'd have had to leave home to be a Land Girl, of course, which I wouldn't have welcomed.

Working at Swynnerton involved a 17-mile journey to work every day. We'd catch a bus up to Burslem Station, then a train to Cold Meece, the passenger-only railway line that was specially built to transport the girls to and from Swynnerton. Then another bus would take us to Swynnerton. I didn't know what to expect, working there, so I wasn't too worried about the work. I didn't know at the time, but the worst thing would come later on – working with the yellow powder.

That first day we all went into a big room to be issued with our security passes. 'Bring these every time you come to work or else you won't be allowed in,' we were warned. And sure enough, you'd be searched going in and even when you were just coming off your shift. When you started work, you went into a big changing room, took your normal clothes off and put on trousers and an overall. Your hair had to be completely covered, no clips or jewellery, so you'd put it up with a bit of elastic. Ciggies and matches were forbidden, too: you had a little hut where you could leave them while you worked. One spark could set things off. We were working with danger.

It was lonely at first in this big place with hundreds and hundreds of women working in different sections. Freda wasn't around either — she hadn't been called up for munitions because she had her job at Johnsons. As an under-18, they started me as what they called a 'first filler of bullets' on a production line. There was a first filler, then a second filler, then a third person to put the cap on.

You worked behind a little protective screen. You had a tray of bullets to fill. First, you put the bullet in; you filled it up with grey gunpowder. A machine came down and pressed it. Then, when your tray was full, it went off to the second filler. You had to wear a turban to completely protect your hair, but there was nothing to protect your hands from the powder you were using, though you could try to wash it off.

We weren't supposed to talk about Swynnerton or what we did there. The powder, even with the turban on, would still have a nasty habit of discolouring your hair. I was in Boots one day and this woman said to me: 'Ooh love, have you dyed your hair and it's gone wrong?' But you couldn't say: 'Oh, I work in munitions.' You'd just smile and say nothing.

Once I turned 18 at the end of 1942, they decided to move me to another section. This time I was filling smoke bombs. Gunpowder was poured into a big machine then the machine would deposit a set amount of gunpowder into a smoke bomb casing. My job was to take one smoke bomb at a time, put it at the bottom of the machine full of gunpowder, then operate the machine to pour the set amount of gunpowder into the casing. Then it was passed on to the next section for someone to put the next filling in.

It was very repetitive work, and you were often moved around the section. I remember one woman, she got so tired, she kept falling off her stool. Little things like that would happen and you'd all start laughing, of course. Anything to break the monotony, which would usually alert the supervisor. I can still hear her now: 'Seeing as you're laughing, you can get on that machine NOW!'

The production line never stopped. As one shift went off – I was on Blue Shift – so another shift was coming on; it was all going 24 hours a day, every day. The factory was so huge, I had a friend working there, also on Blue Shift, but we never saw each other at work. That's how enormous the place was.

The pay was very good: £6 and six shillings for one week of nights. You'd turn the money over to your parents. I got six shillings back from my mother, and she also paid for my clothes out of the £6. So I could still be Fancy Pants.

There were a lot of married women working with me. They'd say things and laugh amongst themselves and I didn't always understand what they were on about. One day, one woman said: 'I'm skint, I've got no money. I'll have to go down Broad's Corner.' This was said as a joke, of course, because a few women laughed. But I didn't realise what she meant.

That time, I went home after work and repeated it. 'Where's Broad's Corner, Mum?' I asked, all innocence.

'DON'T YOU DARE SAY THINGS LIKE THAT!' was my mother's response. 'That's where all the prostitutes go!'

I didn't say any more, but it took a bit of asking around to find out what a prostitute was. A lot of us didn't know anything then. But not everyone at work was that innocent. Two of the girls in my section got pregnant while they were working at Swynnerton; the rumour was they'd been going with American soldiers. They didn't sack the girls or anything like that. The girls went off and had the babies and the babies went into homes. Then the girls came back to work. One day, one of the girls I was talking to in the canteen told me: 'This American told me you didn't get pregnant if you did it standing up.'

I definitely wasn't the only innocent one.

My brothers were always warning me too: 'If I catch you with an American, you'll be for it!' warned my brother George. He was just being a big brother, of course, but all this talk was enough to make me see I'd be better off steering clear of the American soldiers. They seemed to be everywhere by then and they would try to chat you up at any opportunity.

'Can I walk you home?' one said to me as I was walking home one day.

I already had my answer off pat.

'Oh, I'm spoken for.'

But they wouldn't give up.

'I'm only taking you home.'

I'd say: 'Have you got a girlfriend? Would you like someone taking her home?'

'Oh no.'

'Well, there you are then.'

That usually worked!

But with so many men away from home, there were temptations for some women. A woman living near us had a baby while her husband was away in the war, but he forgave her. That sort of thing went on all the time. And with three young men in our house, girls were always after them. If a girl fancied one of them, they wouldn't hold back, they'd just knock on our door and say: 'Oh, can I have a drink of water, please?'

I'd know what that meant.

'JOE!' I'd yell at my brother. 'It's for you!'

The worst part of it all was working nights. And the blackout. At work, you'd be closed in completely, no light other than inside the section where we worked. You couldn't show even the tiniest chink of light. You couldn't even open a door like you could during the day, in case the light revealed where we were to the planes above. Even if you went to the toilet, someone went with you just to be on the safe side. The Germans knew Swynnerton was there – and what it was. They tried hard to get us, but they kept missing.

Everywhere was closed in and dark during those night shifts while we worked away. When we'd hear the air raid siren, our first reaction would be: 'Oh good, we can stop work and have a rest.' And off we'd go, trooping down to the big underground shelter; everyone down there until the all-clear. You weren't allowed to take anything in with you, though. So we'd chat amongst ourselves, and very often, someone would start singing – and we'd all join in.

At home, if the sirens went off, we had our air raid shelter in the garden. I'd get home from my early shift in the late afternoon and I'd be so tired, I'd fall asleep before you could bat your eye, I was that exhausted. And wouldn't you know, the next thing – or it seemed like that – the sirens would go off. Just when I'd gone and fallen asleep in my bed.

'SIRENS'VE GONE!' someone would yell. And so I'd stumble into the Anderson shelter and go back to sleep again until the all-clear. Then I'd go back to bed again. You got used to that topsy-turvy life.

About nine months before the end of the war, they moved me to a section where you worked filling paper tubes with explosive material in the form of a yellow powder. You weren't directly told that this yellow powder was going to affect your skin but you were only allowed to work in that section for a week, then you had a month off the section. And you were tested by a doctor after the week working with it. He tested your breathing, your blood pressure, so it was obvious to everyone that they knew it was a dangerous, toxic chemical.

Yet the effect of working with it didn't happen all at once. It was gradual. Your skin, your hands, your hair, everything gradually went yellow. Even the whites of your eyes would be yellow. So then you'd go off the section and work elsewhere. Then, just when it was beginning to fade, the month would be up and you'd have to go back to the section to work with it again.

We were given thick pancake Max Factor 'cake' makeup for our faces because it was supposed to protect your skin when you worked with the yellow powder. But it didn't protect your skin. Some of the girls would try to sneak the pancake makeup home, even though we were searched coming out every shift. They just wound up looking like ghosts with all the pancake on, that was all.

One good thing we did have was the dinner. They made sure we had a good feed in the canteen. Now and again, you'd be sitting in the canteen and someone would shout

'CAKES!' and you'd leap out of your seat and run up to get them to take home. For some reason, you were allowed to take the cakes out of the building.

In the canteen we'd sometimes play cards, or dominoes. At times they'd organise a big dance with the soldiers from the base nearby at Trentham. I loved the quickstep; Glen Miller's 'Chattanooga Choo Choo'. I'd wear a favourite brown and cream dress with inverted pleats with very high heels. Now and again I'd resort to using gravy browning to draw a seam down the middle of the back of my stockings – that meant they were 'fully fashioned'. But not often.

One night I went to a dance in Hanley with my friend Freda and we got chatting to two boys. One called Tom told me he was a blacksmith, which was a reserved occupation, so he didn't get called up. Tom wanted to take me home when the dance finished. 'Oh, our Jim fetches me,' I told him. And that was the end of it.

But Tom wasn't going to give up that easily. A week later I was at the Baths in Burslem – and he turned up there. I don't know how he found out I went there but he was a fantastic diver. And I already knew he was a nice dancer. Tom was eight years older than me but from that point on, we started courting. I'd had one boyfriend before, a friend of my brother Jim, but it had fizzled out. He'd lived in the same street, but he wasn't a dancer.

Tom would meet me at Burslem station when I'd finished the afternoon shift and he'd walk me home. It was a three-mile walk from where he lived to Burslem but we were young. And both dead keen on each other. The first time I met Tom's mother – my future mother-in-law – I'd been working a week with the yellow powder.

'Alice, is there a bit of Chinese in you?' she asked me. And she was serious!

There were a lot of accidents, though I never actually witnessed one. It could be really scary if someone accidentally dropped a detonator. You'd hear the shout: 'DETONATOR'S FALLEN!' When you heard that, everyone had to freeze. No one could move an inch until they'd picked up the detonator.

My friend Doris worked in the cordite section and lost the end of her finger in an accident. But mostly, the bosses kept any bad accidents very secret. But of course, we'd know if there'd been something bad happening because of the security. On one occasion we were all in the changing room when we were told we just had to stay there: we were waiting there for an hour.

'Oh, they're not ready for you yet,' was the official excuse. But afterwards, word got out: there'd been an accident. A bad one.

We did have a good foreman called Albert. He treated the girls nicely. I always used to think that somehow the men in the factory had an easier job than the women, we were on the production line working away all the time; to me it seemed their jobs were a little bit easier. But we'd never say anything, of course. It was incredibly tiring working those shifts week in, week out. I already had poor circulation. You were usually exhausted when you got home.

Everything was strictly monitored and checked. It really was the kind of security where no chances were taken – ever. One day I turned up without my identification card. Stupidly, I'd left it at home. They'd seen me turning up for my shift every day, they knew my face, but they were still obliged to question me: who are your mother and father? That sort of thing. You'd get fed up with it, of course. But there was a war on and nothing could be overlooked. I never left it at home again!

Our families were involved, too. My mum wound up working in munitions at the small arms ammunition factory at Radway Green, in Cheshire, but the work wasn't as dangerous as Swynnerton, though she never told me what she did there. She was a strong woman, my mum. Tom's sister Annie also worked on munitions at Radway Green. Her husband was a PoW for four years; skin and bone when he came home. But at least he came home.

My dad was incredibly brave; he kept up his rescue work in the pits right through the war. He'd go in if there was an explosion and help rescue people. He never said much but those times must have been hell.

Tom would say all the time: 'Let's get married, then you'll be able to leave Swynnerton.' But I didn't want to. With all the difficulties around munitions work, the yellow powder, being exhausted, I was happy living at home. Where we lived, everyone around us had always helped each other out whenever they could. And that never changed at all during the war. We were there for each other; that's why we could get through it.

I carried on working at Swynnerton for three months after the war ended and finally left in August 1945. I did feel a bit sad. I'd made so many friends there – Irene from Leek, Joy from Trentham, we'd loved going dancing together. You do get close when you work together for so long.

Tom and I got married just after I left. By then, I'd gone to work at Kents in Burslem, the place I'd wanted to work at in the first place, this time as a machinist. All my neighbours donated their coupons to get food for the wedding; the rationing was still on. It was all weddings for us. My brother Joe married that September, Jim at Christmas and George the following Easter. 'You've started them all off,' my mum told me.

We went on holiday to Majorca not long afterwards. They were very poor there then. I can remember seeing cars with no doors on. Tom and I lived with his sister for a while, then we found this house, where I still live. The rent was four shillings a week when we moved in but because I didn't have my first daughter, Margaret, until four years after we married, I was able to save. So we bought the house.

After I had Margaret, I worked at Adams' Pottery as a packer, packaging crockery for eight years until Janet arrived. And after I'd had Janet I went back to Kents, making radiants for electric fires until I retired. Tom carried on in the mine as a blacksmith for 50 years. He was 88 when he died.

A few years back, I met a woman who told me she'd received the same letter as I had after the war broke out, informing her she had to work in munitions at Swynnerton. She told me: 'Oh, I did a week and then I got a doctor's note, so I didn't have to stay.' She seemed quite proud of this.

I told her: 'If we'd all done that, we wouldn't have won the war.'

And it was true, wasn't it? Though none of us thought, at the time, that we were doing something important in munitions. You never told anyone outside your family where you worked. The propaganda was all around us: 'EVEN THE WALLS HAVE EARS'. So you kept quiet.

People often say to me: 'Oh, the girls of today wouldn't do

what your generation did in the war.' But I don't agree; I think our granddaughters and their daughters would do it. The difference was, we didn't really know what was happening all around us. We were in the dark – in more ways than one.

People just weren't that well informed then. There were newspapers and radio to tell you about the war, but that was it. We didn't know the truth about what happened at Dunkirk until the war was over. [Wartime censorship meant that the full story of the maritime rescue of 300,000 retreating Allied troops, stranded on the beaches and harbour of Dunkirk in France in May and June 1940, was not revealed in detail to the public at the time.]

I was a bit miffed that we were never formally acknowledged for what we did, but it's over and done with now. My brother George always used to say: 'It wasn't bad and we were all happy.'

And I agree with that.

CHAPTER 10

DOROTHY'S STORY:
THE GOVERNMENT
INSPECTOR

'DAD WAS A MINER. HE EARNED £5 A WEEK.
I WAS GETTING £8 A WEEK'

Dorothy Orwin was born in Barnsley, South Yorkshire, in 1923. A keen grammar school student, war disrupted her studies and at 16 she went to work in a grocery shop. After signing up for munitions work, she was sent on a three-month Government sponsored engineering training course in Sheffield. This led to her wartime role as an Inspector for the Ministry of Defence, testing components and materials used in armoured vehicles for over three years. Dorothy married her teenage sweetheart, Cyril, in 1944 and their son, Brian, was born in August 1945, just a few months after Dorothy's war work ended. Cyril died in 1994. Dorothy still lives in Wombwell, the small West Riding town she has lived in since childhood. This is her story:

My earliest memory is toddling down to the next door

neighbour's garden and pulling all the heads off his flowers. Though for some reason, I didn't get a smack for that!

There were glassworks and paper mills in Barnsley then but mining was the big employer in the area. My father, Leonard, found a good job working for the Wombwell main colliery when I was four, so we moved to Wombwell then – my parents, me and my sister Kathleen or Kay, 18 months older than me. After we moved, four more children, two boys and two girls, were born.

Wombwell has always been a nice town, well looked after. All the men in the area worked in the colliery; there was no sign of any poverty around us. It was definitely a happy childhood, living in a nice little three-bedroom semi, rented from the local council. My father wasn't actually working in the pit, cutting coal; his job was road making for the colliery (the collieries were all privately owned at the time). This road-making job meant he had to hire his own labour: he had someone working with him that he paid out of his own wage.

We had our grandparents around us too. My mother, Harriet Elizabeth, had her parents living in Bramley, just outside Leeds and we saw a lot of them on weekends and holiday times. We'd get a bus first to Leeds, then a tram from Leeds to Bramley. Grandmother Sophia was quite deaf – you had to really shout at her to get through. And she was a very strict Christian. No work at all was permitted on a Sunday. In my gran's house, Sunday lunch was always cooked on Saturday and you ate the meat cold on Sunday.

My dad's parents, in Barnsley, were quite different. Oldroyd, my grandfather, sang in the church choir and Grandma Louisa was a plump, motherly woman, always on the spot, ready to care for her family or help out wherever she could. It was an outdoor life for us children, walking a lot, climbing onto a swing where you'd play for hours – happy times. I was quite naughty; most kids are, aren't they? But we were not allowed to be cheeky. We were taught to say grace before meals, to leave the table properly, be well-mannered.

We were a church-going family: church every Sunday without fail. Everyone did that then. You always wore your Sunday best for church. One aunt was a dressmaker and she made a lot of our clothes: we always had new clothes for Whitsun. In Lancashire and Yorkshire, there is a Whit Monday tradition: a big walk around the town by the local church and chapels. We'd meet up at the church and then we'd all walk along with the local brass band playing. Each chapel carried its own banner. Then we'd troop back to the church for a lovely tea – a highlight of our year, really.

At home, of course, we kids all had our jobs around the house. Lots of washing-up and dusting, cleaning some of the windows, scrubbing the doorsteps. The edge of the step always had to be scoured with a special donkey stone [a scouring block made from pulverised stone, cement, bleach powder and water in different shades of cream, brown or white]. Every doorstep was scoured with the donkey stone – it gave your doorstep a nice decorative finish.

At five, I started at Wombwell Park Street Primary. I was fairly good at school. I passed what was then the 11-plus exam and moved to a secondary grammar school, at Wrath upon Dearne, just a few miles away. I was there from 1935 to 1939. I left school just a few months before my 16th birthday, weeks before war broke out.

I'd enjoyed studying, but with war on the horizon for us,

there didn't seem much point in me staying on with a view to going to university: had I done so, I'd have been the first in our family to go, though my sister Kay was an equally good student. But we all knew that times were about to change – and that things were going to be increasingly difficult.

That summer, we'd all been looking forward to our annual family camping holiday in Penistone, in the foothills of the Pennines. But at the last minute, we had to cancel. Dad couldn't go away because he was suddenly needed by the colliery. They had to keep everything fully manned so we couldn't go. We wouldn't dream of going without him.

To tell the truth, I was a bit frightened that Sunday morning when we heard the news that we were at war. The idea of fighting in our country, being invaded by German soldiers – well, I had a vivid imagination, anyway, so it did affect me, though I kept my thoughts to myself. Yet the fear didn't stay with me – it was more the effect of listening to those sepulchral tones pronouncing: 'We are at war with Germany', that did it for me. And, of course, nothing happened straight away, the phoney war period.

In the beginning, there were no air raids, though the blackout started straight away. We went out and bought black material for the windows: we couldn't use torches at night. And the streetlamps were not lit in certain places. We weren't very far from Sheffield, either. It was already believed that the city would be a target because it had so much heavy industry, including steel and armaments works. But right at the very beginning of the war, no one really knew what was going to happen. So at first, life went on pretty normally.

In 1940 I got a job in a grocery shop. Soon, I was serving people coming in with their ration books (buff coloured for most, green for pregnant women and under-fives, blue for children ages five to 16). That January of 1940, bacon, butter and sugar were rationed, quickly followed by the rationing of virtually everything else you could think of. My years at school doing arithmetic came in very handy in the shop. We didn't have tills then where you just pressed a button to calculate everything; it was all done mentally. You handed the goods over from the counter and you added everything up with a pencil, on the spot.

My dad was still working at the mine – a reserved occupation, of course – so there was still money coming in. But the food shortages got worse as time went on. Yet there were no angry customers coming into the shop, complaining about it all. People just seemed to accept the situation. Of course, some people got a bit peeved about not being able to get as much as they wanted but the main idea of rationing was that everyone had an equal share – that was the theory, anyway. The big cities with docks like Liverpool or London was where the black market [rationed goods being sold off for cash] really thrived, because there was greater opportunity to unload 'black' goods. But in our part of the world, you didn't see or hear much about it.

After we'd closed the shop for the day, we had a fire-watching rota system where we would have to take it in turns to stay on duty in the store for a couple of hours. Fortunately, nothing much happened, but by the summer of 1940, the German planes were flying over Sheffield, and that December there were two terrible bombing raids – the Sheffield blitz – which killed nearly 700 people, demolished thousands of houses and damaged some of the steelworks.

That, of course, brought the truth of war, the devastation

and the heavy losses, home to everyone. We could see the flashes and hear the noise at Wombwell because we were right on the top of a hill, overlooking the valley. So we saw it all happening – from a distance.

Towards the end of 1941, I turned 18. Now I had to register for war work. I definitely didn't want to go into the Forces, so I signed up for munitions. I didn't have a clue what it might entail, of course, so I was entering completely unknown territory. But because I'd gone to grammar school and my studies had been disrupted by war, rather than sending me straight off to work in a factory, I was considered to be suitable for some kind of training. So for 12 weeks I was sent to a government training school in Sheffield. It turned out to be an engineering course where I learned about all kinds of machinery, along with a number of other girls around my own age. The training involved using different machinery; learning how to operate a turning lathe, for instance [a lathe is a machine tool which rotates the workpiece on its axis, so it can carry out other operations like sanding, cutting or drilling] or how to use certain measuring instruments, like callipers [a calliper measures the distance between two opposite sides of an object].

The training school in Sheffield was about 14 miles from home and it was a bit tiring because it meant shift work, which I wasn't used to. I didn't like working at night, climbing the stairs at Sheffield station to get a train at 8pm, finishing training at 6am to get back on the train home. It was a bit scary for me sometimes, after the big bombing in Sheffield, though the December raids turned out to be the worst of it. [The bombing raids on Sheffield ended in July 1942.] But the training was only for a matter of weeks – and

I didn't know it then, but I wouldn't be asked to do shift work again.

At the end of the training period, all the trainees in my group had to sit a written exam. From that they chose three or four girls to work as armaments inspectors, checking items before they were assembled – and one of those girls was me. I was told the work would be as an inspector with the Ministry of Defence. Our section was called IFV, the Inspectorate of Fighting Vehicles. That meant inspecting tanks, Bren gun carriers, anything that was armoured.

When I came home that evening and told my parents all about what I'd be doing, they thought it was wonderful, they were quite proud of me. And they were pleased to hear it didn't involve any more shift work – but perhaps not quite as relieved as I was! What we didn't realise at that point was that my earnings would be so good, I'd eventually be earning more than my dad – he got £5 a week, I wound up earning £8 a week sometimes.

My very first job as an inspector involved a posting away from home to Nottinghamshire. I was sent to work at the Ordnance Depot at Chilwell, an area between Derby and Nottingham. During the First World War, Chilwell had been the country's most productive shell filling factory, but just weeks before WW1 ended, there had been a shocking explosion at Chilwell, killing over 134 people and injuring 250. After that it became a storage facility.

So there I was with my new job, away from home for the first time. Another girl, Joan Pollard, was also starting the same work for the Inspectorate, so we wound up sharing a room in a boarding house in Long Eaton, the first time I'd ever shared a room with a total stranger. But as it turned out, we got on really well, so well that on weekends, we didn't go home. We'd go into Derby or Nottingham for a look round. We had to pay for the accommodation but we did get an allowance for it in our wages.

That first day at the Chilwell Depot involved a medical inspection. About a dozen girls my age were all ushered into a little ante-room and told to undress to our knickers. Then, after a bit, the door opened. Naturally, we were expecting a doctor or a nurse but to our surprise, in walked a squaddie [an ordinary soldier]. He ignored us – all desperately trying to cover ourselves – walked up, whistling, to a cupboard, opened it, took out a sweeping brush, and calmly walked out. As you'd imagine, we all fell about laughing once he'd gone. We reckoned he did it deliberately every time a new group of girls came for their medical!

Then came the medical itself, lots of tapping and checking nose, ears, throat – more a basic check-up than a thorough medical. We didn't question it, or its purpose. After that, we were interviewed individually by an officer. He talked to us about what we'd already learned in our training. Then our duties were explained to us. We'd be checking all manner of items: the general idea of the training course had been to get us used to using different types of machinery which would help us do our work, inspecting and checking different parts or components that were being assembled. In simple terms, we'd be testing these things to make sure they'd been made correctly.

Chilwell seemed like a huge place. It seemed to be run by the Army and the ATS: very big tanks were being tested there. But as a novice inspector, you started out small. My first job at Chilwell involved inspecting small items like rubber circles.

You had to measure the item to make sure that the rubber was the correct width and was suitable to fit onto the component it was made for. We worked with drawings, to make sure the items matched the measurements on the drawings. My maths lessons at school had not been wasted, after all.

I didn't spend very long at Chilwell, as it turned out. I worked there for a matter of weeks. After that, I travelled from home to wherever they sent me: no more boarding house. My work after Chilwell involved tanks, working for a company called Toledo Woodhead Springs. They made springs for heavy automated transport, coil springs and leaf springs that fit onto the body of the tank. These had to be tested for hardness, to make sure they could withstand the wear and tear on the tank. We'd also have to test the hardness of the sprocket wheels before they were fitted onto the tank. I was also testing the armour plating, the hardened sheets of steel used for the body of the tank.

I'd look through a microscope to measure the imprint of a steel ball, to work out how hard or soft the imprint was making the steel. If the imprint was too big, it meant the steel was too soft and it wouldn't work. If the imprint was too small, the steel was too hard and it would have cracked. So there was a tolerance you had to abide by.

The odd thing is, at the time I was doing the work, I just got on with it. I didn't focus on what it all really meant, making sure these huge tanks were safely assembled and could be sent out for combat in the front line. It was only afterwards that the significance of what I had been doing really sunk in. Get it wrong just once and lives were at risk. Had I made a mistake and passed steel that was not worthy of its job, it could have been chaos. For example, if a steel plate on the body of a tank had been over-treated it would be too hard, and a shell would shatter it. Too soft, and the shell would go through it. Quite often, the steel plates had to be rejected. Then they were re-heated and quenched again to get the correct hardness.

You had to be a very careful worker. The men would swing a huge steel plate onto the big table to be examined. Another worker would use a grinder to grind a smooth place in one corner of the plate – and I would have to measure the imprint using a Brinell, a bit like a small version of a telescope, which measured the imprint.

The work was absorbing and while I had to work in some very noisy places, I wasn't on a production line; I worked on my own. I'd be based in an office, so whenever something was ready to be inspected, I'd have to go out onto the shop floor and do my work. I didn't wear a uniform as such, just a white overall. My hair didn't have to be covered either; I wasn't actually working near any machinery.

The pay was very good. I gave some to my mother and managed to save some, too. My older sister, Kay, wound up working in a similar job in an aircraft inspection department just outside Leeds at the Avro company, where they made planes like Lancaster Bombers. Then she moved and went to work for the NAAFI. She wound up travelling all over the place, and was one of the first women to land with the troops at Anzio in Italy in 1944.

I vividly remember walking on the moors one day, thinking of Kathleen and picking a sprig of heather to send to her in Italy, a small reminder of home. She told me later it wound up being displayed in the Officers' Mess and they nicknamed the NAAFI canteen 'The Heather Club'.

Others in my family were involved in war work, too. My younger sisters, Elsie and Audrey, were too young for war work but my brother Frank, a few years younger than me, worked in a steel factory in Sheffield before joining the RAF as a driver. My other younger brother, Leslie, eventually joined the Navy. As for me, I'd already started courting even before I went to work at Chilwell. I'd met my husband to be, Cyril, at the local Methodist chapel in Wombwell in 1940. He was three years older than me and in a reserved occupation, working as a joiner in the local colliery.

After my stint at Toledo, I was sent to work at a place called Shorter Process in Attercliffe, a suburb of Sheffield. I actually thought the name meant 'short process' at first – but it turned out the boss was named Shorter. At Shorter they made sprocket wheels for armoured vehicles. A labourer would lift the sprockets, two feet across, and I would have to press a testing machine, called a Firth hardometer, which used a ball or a diamond pressed into the metal to measure the imprint: a diamond is harder than steel.

I worked there for a few months, and the last place I worked at was called Firth Browns, a large engineering factory in Monk Bretton, just outside Barnsley. There were a lot of other women working there. There was also an underground firing range where they'd test the sheets of metal used in the tanks by firing live ammunition at them. I didn't take part in the firing; my role was to measure the metal sheets afterwards.

I never really talked about what I was doing with anyone other than my family and Cyril. And the rather solitary nature of my work meant I never really made friends in the places where I worked. I wasn't actually involved in a process where something was being made; I was being called in to test. I did have one friend from Wombwell, who had originally worked in the grocer's shop with me, Sadie Green. She was sent off to work in an aircraft factory repairing planes. She worked on magnetos, the electrical generators used in aviation piston engines, repairing and fitting new ones.

You had no uniform as such in the Inspectorate but you were given an Inspectorate of Fighting Vehicles badge as a security pass. It was circular with a brown outer ring, then a red ring; the rest was green with a side view of a silver tank. And it had a number on the back: mine was 711.

I carried on working for the Inspectorate through the war, even after Cyril and I got married in July 1944 at the church where I'm still a member, St Mary's at Wombwell. It was just a few months before my 21st. The war wasn't yet over, but we felt it was the right time, as we'd been courting for nearly four years.

The local photographer turned up for our wedding photo – only to discover that he had no film for his camera, thanks to war shortages. Two weeks later, he got in touch to say he'd managed to get some film. Of course, my bouquet had had its day, but then another girl I knew, who had just been married, had kept her bouquet. Yet for some reason, the photos he took weren't any good. I didn't like them. In the end, I chucked them away.

After the wedding, we moved in with Cyril's father. Then came a surprise out of the blue: Cyril was being posted down to London, to help work on bomb damaged properties. So off he went, and I went back to my parents. Not long after he'd gone to London, I wrote to Cyril with some good news:

we were expecting our first child, Brian. And, of course, by the New Year, the war was gradually drawing to a close. I carried on working until just before VE Day, in May 1945, and Brian was born in August that year, a real Victory baby. But it wasn't until the end of the year that Cyril finally came back home to Wombwell. Then we managed to get a house of our own: I've lived in it ever since.

Cyril carried on working as a joiner for a private building firm. I did go back to work after Brian started school; first I went back to the grocery shop in Wombwell for 10 years, then I worked in the invoice department at a firm called Newton Chambers in Chapeltown for 12 years. They made Izal toilet rolls. By the time I left Newton Chambers, Brian had joined the Merchant Navy and was working as an engineer officer on the *QE2*. Sadly, Brian died of cancer in 1992.

I enjoyed my war work; there was a real sense of responsibility in knowing you had to be one hundred percent accurate every time you checked something, but I do consider myself very fortunate to have been given that job when you consider the dangers and difficulties involved in the factories so many munitions women worked in.

It didn't really strike me, over the years, that there was very little recognition for the munitions workers but now, of course, you can see that we should have had some kind of recognition for our work. I think a lot of us are like me, not angry or militant about being 'invisible' because we just got on with it, like everyone else we knew.

Yet somehow I don't think today's generation could do what the munitions women did in wartime. They seem to think too much about themselves these days. And I'm not sure either that people nowadays are as patriotic as we were during the war. The world is so different now, everything is 'must have'.

In the war, there was nothing to 'have' but your relationships with the people around you; your family, your friends, were somehow much closer. There was a more genuine neighbourliness then: people cared about people.

'Through the mud and the blood to the green fields beyond' is the Royal Tank Regiment's unofficial motto. That saying sums it all up for me: it gives me a sense of who I was, what I was doing during the war.

And where I was going.

CHAPTER 11

IRIS'S STORY: THE GIRL ON THE BICYCLE

'WE WERE SAVING THE LIVES OF OUR OWN TROOPS'

Iris Aplin was born in 1923 in Market Drayton, Shropshire. Her parents had worked in service when they were young, so when Iris left school, she followed in their footsteps, working as a kitchen maid in various big houses until war broke out. In 1941, age 18, she went to work in munitions at ROF Swynnerton. Iris worked at Swynnerton for four years in many different sections, often working with highly explosive material, filling smoke bombs and assembling detonators and fuses. In 1948, she married her wartime fiancé Bob Aplin and moved down to Honiton, Devon to raise their family. Bob died in 1996. Iris has two daughters, five grandchildren and four great grandchildren. This is her story.

There's a link in my past to the TV show, *Downton Abbey*. My father, Charles Merrifield Young, worked in service. He was a groom in the stables at Highclere Castle, the big country estate on the Berkshire/Hampshire border, now known all over the world as the setting for TV's fictional Downton Abbey.

My mother, Amelia, worked in service too as a nursery nurse. My parents did their courting in Highclere village – and they were married there in 1920.

My mother's parents lived in Market Drayton, so my parents lived with them after they married; I was born there three years later. Dad had always worked with horses but after WW1, he had difficulty finding work – until he found a job as head pigman for a farming company at Kinsey Heath in Cheshire. Then we moved there.

My earliest memory of Kinsey Heath was getting the water from the pump in the garden and going shopping in a horse drawn cart to Market Drayton, where we'd go to visit Annie Chidlow, my grandmother.

Annie scrubbed people's steps for sixpence a time. She made a good living out of it. In the end she owned three houses in the area, rented two out and kept one for herself.

We lived in a little rented cottage with my baby sister, Ruth, who arrived three and a half years after me. I hated the cottage. At night, there were insects like cockroaches, horrible black things that had a nasty habit of crawling into your clothes. One got down my back one morning as I was on the way to school – poor mum, I was yelling the place down. But we were happy living there, in the countryside. Ruth and I used to have to walk a long way to school because the houses in the area were quite a distance apart.

One day my mum put up some curtains in the window of the cottage and the owner of the cottage decided he didn't like the curtains – they weren't good enough for him and he didn't like the colour either. His daughter lived next door to us so my mum soon got the message 'alter those curtains'. But my mum wasn't having any of that.

A week later we went to see an old house nearby at a place called Coxbank, about two miles from the cottage. Mum knew the owner, who lived in Crewe, and the owner agreed to rent it to us. I was really pleased. It was a beautiful old house – eventually it had to have a thatched roof because we had to galvanise it because of the incendiary bombs in the war, though in the end, we didn't get hit.

I was the biggest girl in my class. I was a well-made girl, as they used to say, mainly because we ate well. My parents grew their own veg and fruit and we even had plum trees in the garden. Mum and dad also kept chickens and our own pig. Me and Ruth used to help dad outside sometimes; we all worked together.

I didn't like the school headmaster, Mr Coffin. He was always going on about
'the old days', the Victorian era, when everyone slept in one bed, parents and children.

He was telling my class about it one day and he managed to upset me no end.

'Iris was in the middle,' he told the kids.

'And when she turned over, they all had to turn over.'

Everyone laughed but I never ever forgave him for that comment.

I was not a shiner at school. At 12 I was sent to a secondary school in Audlem, Cheshire. I definitely didn't want to go to the big posh grammar school. We had cooking, washing, gardening lessons. I came top of cooking, making a Swiss Roll. But I left at 14.

Round our way the only kind of work available for girls was in service. You just followed everyone else into it. Straight away, I found a job as a kitchen maid at the old Rectory in Adderley, Shropshire, a very big house but just me and the cook doing almost everything. The pay was five shillings a week and I went to live in, up in the attic.

Downstairs I can remember seeing rows of brass bells for every room, so there must have been a lot of servants there at one stage. But now it was just a parson and his wife – with a cleaning lady coming in to help out. They were nice people but it was hard work, day in, day out, with only Sundays off.

A year later I switched jobs to a place in Market Drayton called The Grove Hall, same kind of job, still sleeping up in the attic, and then I moved jobs again, to another big house in Market Drayton. I worked for a couple, a Wing Commander and his wife.

By this time, the war had broken out. My dad, who'd been in the first war, said 'Oh no, not again' though as a 16-year-old, I didn't quite understand it all. But it wasn't very long before I understood what war really meant. I was at a dance one night at Audlem; Dad worked on the door sometimes. One of the local boys was at the dance in his uniform, off on leave from the Army. Two weeks later, we heard he'd been killed. That struck home to everyone.

I wasn't called up. What happened was that my Wing Commander boss and his wife suddenly had to move from Market Drayton to Reading in Berkshire. They told me they needed to find someone to work for them as a servant, and much to their surprise, I said I'd be willing to go with them. At 17, I was happy to move away from home. I liked my employers. The lady of the house was bedridden, so I got to do different things helping her and she was a very pleasant person to work for.

My parents' attitude was 'well, we had to do it for work' so they were fine about me moving to Reading. I was still sleeping up in the attic, but I was now that much older and starting to understand that because of the war, there were other things I could do for work. Girls like me from rural families who'd only ever known service welcomed the fact that there were opportunities for us to do something different for the first time – and earn a bit more money. So I left and moved back home.

In the spring of 1941, I turned 18. My best friend Mary, who was also working in service, was of the same mind as me: we'd be better off doing war work. We'd heard there were lots of jobs for young girls at Swynnerton in the munitions. At first, I'd wanted to go into the ATS. But when I broached the subject, my parents wouldn't hear of it.

'Not a place for a woman, Iris,' my mother said.

And so Mary and I got on our bikes at Coxbank one day and pedalled furiously over the mile and a half to Audlem, where we were able to leave our bikes with friends. Then we boarded the bus for the journey to Swynnerton, over 30 miles away. It took ages but the time flew by, mainly because we were singing all the way there. We sang war tunes like 'I'll be with you in apple blossom time'. Other people on the bus joined in. That's what you did then.

At Swynnerton, they took down our details and told us we'd need to have a medical first –this would be done by our local doctor. So they handed us both a receipt which we were told to hand over to our doctor. Then, after the medical, the doctor handed us our reports. Within a week or so, we were on our way to Swynnerton again to get ready for our first ever shift.

Swynnerton was huge, a world of its own with strict security and lots of rules and regulations. Mary and I were put on Group Six to start with. That was where they made the smoke bombs. The overalls were white and you had your clock-in number stencilled on the back. My turban was green – to denote Green Shift – and you also had to change into the regulation shoes they provided. I take a size eight, so they had a job finding a pair. They were brown shoes with a crêpe sole and an arrow on the toecap.

And so I found myself standing on a big production line in a large area with double doors at the bottom and the top, and an annexe attached at each end of the shop so the truckers could get the stuff in.

What I didn't realise beforehand was the fact that you were going to be on your feet while you worked. The shell cases came along on the production line and about 25 of us all had to do exactly the same thing; fit a little upside-down adaptor – like a little wheel – on the top of the shell case. That was our part of the process. Then it was passed down the line to another place for the next step in assembling the smoke bombs.

As we worked, a woman would be walking past us, keeping an eye on everything we did. Her armband said 'Chief Inspector of Armaments' and she was there to check (and double check) that the work was being done properly. No-one was taking any chances.

The head or boss of the shop was called a 'Blue Band'; his name was Jack. Jack was good to us girls, but he did have one or two funny ways. For instance, we frequently sang while we worked to cheer ourselves up and relieve the monotony. But if you were on night shift on a Sunday, come midnight, if we were singing a hymn – which we often did – Jack would always say 'Sunday's gone, girls' which was code for 'no more hymns, please'.

Jack was also known to use colourful language sometimes if something went wrong – but that didn't happen very often. At times a smoke bomb would be filled too full; they were only little things, but if that happened, it would put the machine out a bit.

The girls liked Jack because he had the right attitude: the priority was to get the work done. So he was respected by us. And he always allowed us that little bit of time before we took our hour's break, the point when we had to make sure that everything was safe, the machines switched off, before we could leave the floor for the canteen. It was just a matter of a minute or two. But it meant a lot to us.

Jack was also very good to the girls when we worked on nights. There'd be times when the shop got too misty – with all the grey powder that was being used – so he would get us all to stand there, with the lights out, while the doors were opened very briefly for a bit of fresh air. (At night, of course, the total blackout meant all the shop doors had to remain closed so not a chink of light could get in). We'd all stand there for less than a minute, then he'd shout 'close the doors!' and then the lights could go back on. Jack was a very understanding man.

At home, my job at Swynnerton had another effect on the family finances. I can remember going home with my first pay. Dad was sitting at the table and I put my wage packet down in front of him.

'Look at that dad, sixteen shillings,' I said, pleased as punch with myself.

Dad just picked it up, looked at it and said: 'Right. That's the end of me working on the farm.' His daughter was earning more than him! So he found a munitions job, as a fitter's mate at Ternhill airfield.

War work meant that I had real money for the first time. Half my pay packet went to my mother, the rest was mine. The old people in the village would sell me and Mary their clothes coupons. As a result, we had more clothes to wear when we went out dancing.

After about three months on Group Six, Mary and I were moved. She went to Bullets and I moved to 7C, working on detonators and boosters, making the components to put the fuses for the bombs together.

This kind of work was dangerous. We were working with highly explosive material, TNT. You worked with a tray filled with little handles for the fuses, and each one had to be filled with the explosive material, the TNT. You had to fill each one by hand. You used a little tiny scooper. You would scoop a small amount of the yellow TNT powder from a metal container and place it on each handle. Then there was a big machine that pressed it all down. At that point, you had to walk away and stand behind a protective glass screen while it was being pressed – in case it exploded.

At times you wore gloves to do the work but not always. When you finished work for the day, your hands were stained yellow by the powder. If you blew your nose, your hanky would be yellow. Some of the yellow would come off if you washed your hands, but not all of it. And a lot depended on what type of skin you had. For some girls, the yellow powder didn't suit their skin at all and they had all sorts of allergies and skin problems with the TNT. Some girls' skin problems were so bad, they would have weeping sores on their faces, so they wore a bib permanently – and then they wound up being off work for good.

One of the things I did when I saw the effect the yellow powder was having on other girls was to go home and rub my face with peroxide: that neutralised it to an extent. But I was very lucky; for some reason I didn't have serious skin problems working with the yellow powder. I did have a blonde bit in my hair where the turban didn't fit properly, but that was the only effect it had on me.

The early shift meant a pre-dawn start. Up at 4.30am, pedal like mad to Audlem, then a 5.30am bus ride to Swynnerton to get there in time for the 8am start.

On the later shift, I'd be home at midnight. But if my hours permitted, I'd still get out and help Dad in the garden, or get the tea for Mum. By now, she was working too as a Pearl insurance collector. She got the job from a neighbour who'd had the round but he had to go in the Army. So Mum did it for him through the war.

I used to work in the garden sometimes in my skirt and bra; that used to give the neighbours a good laugh. There was an Army camp near us at Adderley and they had a habit of practising their bugles in a nearby field. Sometimes at night you couldn't sleep for the practising!

But if it was tiring travelling backwards and forwards all the time it was still better to live at home than being sent off to live one of the big hostels they built specially for the munitions workers.

A couple of the girls at work stayed at one of the hostels

and they were always saying how much they didn't like it. Though from what they said, it was all very organised. I felt quite sorry for them. They were obviously very homesick.

But that was how it was in munitions: the women working with me came from all over the country, places like Scotland and County Durham. It was tough without their family close by. Or having their husbands or sweethearts posted abroad. And the food they got at the hostel wasn't what some of them were used to. I can remember some of the Scottish girls saying they couldn't stand the celery that was usually served on a Sunday evening; they used to give it to the English girls. And the Scottish girls thought some of the English girls were 'fast' because they used to talk about drinking alcohol or going to pubs.

There were laughs too, though. In the canteen, the girls who packed the boxes of bullets used to talk about the little notes they would put into the boxes for the soldiers or sailors.

'Keep 'em on the run' they'd scribble to these troops. Or 'We all love you and we're here for you'. They'd even include their address, just in case. I don't know if it's true or not, but I heard that one or two girls even wound up marrying men who'd got the letters. There was a lot of talk about that sort of thing going on.

For us, the main thing was we were doing a job, saving the lives of our own troops. We didn't think about the enemy or what they might be doing. In my early days at Swynnerton, us girls would sing the regimental marching songs like 'Colonel Bogey' – they all had rude words – yet gradually, as we got the reports of how many troops were being killed, we stopped singing those songs. It just didn't seem right any more.

Then I was moved again to another shop in 7C, making tracer bullets for the pilots to use. (Tracers are slow burning units, used as markers for targets, built with a small pyrotechnic charge in the base). The tracers were filled with something that looked like mouse droppings called strontium nitrate.

As dangerous as the work was, I never actually saw any accidents. But we certainly knew about them. Sometimes you'd hear an explosion – and, of course, there'd be talk about it amongst the girls afterwards. We all heard about one very big accident that took place at the burning ground, which was to the north of the Aycliffe, far away from the factory buildings. (The burning ground was the area where sub- standard explosive material used to assemble the ammunition could be safely destroyed). We had no idea what went wrong that time. But it was a very serious accident.

It was great to have money to spend on your days off but thanks to one of the older women in the shop, I realised there was a chance to save money too. This older lady had worked in a pottery factory before the war and one day, a group of us got chatting about going to dances and buying new clothes when the lady looked up and said: 'You young people won't 'alf be sorry one day. You're earning big money. You should be putting some away for when the war's over'.

That stuck in my mind. And then I learned you could save while you worked. They'd take so much out of your wages every month and you'd get a savings certificate in your pay packet, showing you what you'd saved up. I started out saving five shillings a week, then more. In the end, I came out with £300. Many other girls did the same.

I did have good reason to save, mind you. By the last year

of the war, I got engaged. I met William George Aplin, 'Bob', in July 1941. It was funny how we met, really. I was walking home from a whist drive with my mother and two friends in Audlem and on the other side of the road, this soldier came along. His shoes were making a funny clicking noise and we all started laughing. Then he crossed the road.

'What's so funny?' he wanted to know.

'It's your boots,' my mum said. Then he told us he was going back to camp at Adderley.

'Is anyone going my way?' he quizzed us, looking straight at me.

My mum didn't mince words.

'She's the only single one,' she said, gesturing at me, having spotted his look.

I decided I liked the look of him. He was clean and he looked happy – he'd definitely had a beer – so we walked off together towards Adderley. On the way I heard all about Drake's Drum and how it had been returned to Plymouth. Bob was with the Devonshire Regiment. He'd been in hospital with a bad knee and because he'd stayed in hospital for more than three weeks, he'd been transferred to another Regiment, The Loyals.

Halfway back to his camp, it was agreed. We'd meet up the next night.

I liked this soldier. There was no nonsense about him; he was 25, seven years older than me, and he was more or less my first boyfriend. My mum liked him because he was a tidy sort of person. So we saw each other for all of three weeks before he was sent away, up to Scotland. Bob wrote me from Scotland to say he was going abroad and it was not fair for us to stay attached.

'You could meet someone else,' he wrote.

I didn't reply. I knew what he meant. Other girls at work met young men like this and then they'd suddenly be posted overseas. But I kept that letter – I've still got it to this day.

That was in the August. The following January in 1942, just after Christmas, a card came from Bob wishing me a Happy New Year. There was a PS at the bottom: 'If you ever find time to write, this Post Office address in Burma will find me.'

Guess who sat down and wrote? And we wrote throughout his time in Burma.

By the time I was 22, just before the war ended, he managed to get a month's leave. Bob came home and asked me to marry him. So by the time he went back we were officially engaged. And then the war ended.

Three weeks after VE Day, we were all out of Swynnerton. I was among the first to finish there because I'd been travelling in on the outlying buses. It was over for us. So we had to go back to the land – and start again.

I went back into service, at Moss Hall in Audlem, a manor house overlooking the Shropshire Union Canal. I was cleaning and cooking. But no more living in.

Bob was a stonemason. He didn't like to talk about Burma after he got back. There'd been many times when he couldn't write to me because he was behind the lines and his brother, who was a medic and also out in Burma, knew this and would write to me to say: 'keep writing, Iris'. So I never really knew what it had been like for Bob, how bad it was.

We were married at Addenham Parish Church in March 1948. Bob got his old job back in Honiton, Devon, so afterwards we moved down there straight away because we'd found a house to live in. On honeymoon, I got caught. My eldest daughter Amelia arrived late in 1948, then Josephine in 1953. After Jo was born, I went back to work in a secondary school canteen for 27 years. Bob carried on working as a stone mason until he retired. He was 80 when he died in 1996.

There should be some recognition. We did give up all our time to it, working those shifts for four years. Everyone you worked with mixed in, helped each other out. We didn't think about the danger. But it wasn't like today where people stand off: the comradeship us girls had then just isn't there now. Though sometimes I think women today would do the same as us if they had to. I'm sure they would because they would think about their country.

I often wondered afterwards if those bombs we helped make did their job and many years later I asked one of Bob's friends, who'd also been in the Forces, about it. I was told: 'oh yes, they definitely did their job'. We knew we'd helped. But it was still good to hear it all those years later.

CHAPTER 12

THE FACTORIES

While the day-to-day working life for Britain's Bomb Girls followed a similar pattern across the country, the history of the munitions sites varies from area to area.

The new Royal Ordnance factories carried out much of the wartime production of munitions. Yet the existing, privately owned factories and works, adapted to manufacture munitions for the war effort for the Government, were already employing young women before the start of WW2.

In some cases, these women finished their munitions work when war ended yet continued to work for the same company when the factory returned to its peacetime activity. (Ivy Gardiner did this briefly at Port Sunlight before she got married).

While many of Britain's Bomb Girls worked in the three largest filling factories at Bridgend, Swynnerton and Aycliffe, detailed record keeping in wartime for the other munitions factories located around the country has not proved extensive, mainly because of the secretive nature of munitions work. So it is not possible to give a more detailed overview of all the munitions sites.

Here is some additional information on the various munitions factories where the women interviewed in this book worked.

BRIDGEND, GLAMORGAN
(BETTY'S STORY)

The Bridgend Arsenal was chosen as a filling factory for a variety of reasons: geographically, it was remote from the areas of the country most vulnerable to enemy bombing, situated in a damp and misty area, further protecting it from attack.

It was also in an area of economic deprivation, providing local employment for thousands. Moreover, in terms of logistics, it had good railway links and was also relatively close to important areas: the huge South Wales coalfield and the ports of Swansea and Cardiff.

The first phase of the construction of the Bridgend Arsenal started in 1939, though it was not until March 1940 that the factory was fully operative. The 1,000-acre site was divided into individual production groups, operating as self-contained units dealing exclusively with one kind of part for the weaponry assembly or other activity. (As in all other newer factories working with explosive material, these groupings reduced the risk of toxicity and explosion).

The arrangement of these groups was complex: it had to take into account the storage of highly explosive powders and weapons parts, the way they could be moved around the site to the appropriate workshops, the manufacture or filling of components and their storage when they were ready to be exported to other munitions factories for assembly.

Buildings on one side of each group had offices and changing rooms. In the centre of each group were the process workshops where the women worked. Each group was serviced by railway lines for transportation of empty and completed parts.

The main part of the Bridgend complex, covering about 500 acres, was to the south of the site, named the Waterton area. The main administration of the Bridgend site was run from here.

Waterton also included the Pellet Group where Betty worked, the area where yellow powder was fitted into circular pellets. Near the main entrance at Waterton was the Textiles section, where Betty worked until she turned 18. Waterton also contained the biggest canteen in the Arsenal which prepared meals and food to be served in the smaller, satellite canteen in the various groups. This canteen also had a special stage for entertainments. There was even a separate dining room with waitress service for high ranking employees.

The northern site, Brackla, also around 500 acres, contained seven large magazines (concrete storage chambers) in tunnels 50 feet below ground. These tunnels were built to give some protection in the event of bombing as well as insulation from the effects of an accidental explosion. Open railway cuttings led to the underground magazines, this way the dangerous, yet precious cargo could be moved in and out. Brackla also had a number of sub factories, where high explosive mortars for the Army were filled and fine grain gunpowder was produced for time fuzes. (Fuzes are the machinery controlling the detonation of the main charge, usually including a detonator and a booster, initially used in anti–aircraft shells).

The standard working week at Bridgend was 45 hours. A basic shift lasted for seven and a half hours, including 45 minutes for a meal break.

Each shift was designated a colour (red, green and blue) and the shifts rotated weekly, in an afternoon, day and night sequence. Morning shift at Bridgend started at 7am and finished at 2.30pm. Afternoon shift was from 2.45pm to 10.15pm and night shift started at 10.30pm and finished at 6am the next day. Frequently, machines were kept running round the clock, so that workers could pick up where the previous shift had left off and precious production minutes would not be lost.

Most of the women working at Bridgend were in their early twenties although older women in their fifties worked there too. As Betty Nettle recalls so clearly, there were also a number of men employed at Bridgend, many of them older men or those not fit enough for the Forces. (Some women resented their presence, especially if their own husbands had been called up and sent abroad).

The women themselves came from many different areas across the country as well as from outlying areas like Maesteg, Aberdare or Pontypridd. Pay day was on Friday and wage envelopes were hand delivered. Overtime was available, though it would only be paid after a set number of hours had been worked. The overtime rate was 1- 1/3rd hour's pay for the first two hours, 1 1/2 hours pay for any other overtime worked. All hours worked on Sunday were paid at double time, overtime on Saturdays was paid at the normal weekday rates.

Bridgend was a huge munitions enterprise: at its peak in 1942, it employed just over 32,000 people. After WW2, munitions production ceased. Eventually the site became an industrial estate, with most of the original industrial buildings demolished and new factories and homes built.

Official figures show that 17 people (11 men, 6 women) died in accidents at Bridgend from 1941–1945. No official figures exist for the numbers of accidents that took place there.

Given the nature of the work in filling factories, injuries were often loss of fingers, hands or eyes.

22 year old Gwen Obern was injured in an accident at Bridgend in 1940. The accident killed five people and injured 14 others when a box of detonators exploded. Gwen was blinded permanently and lost both hands in the accident — which took place on her third day of work at Bridgend.

Afterwards, Gwen underwent 76 operations, learned Braille and became a leading light of the St Dunstan's organization for the blind. Undaunted by her disabilities, she continues to live in Wales.

THE ROSES OF SWYNNERTON, STAFFORDSHIRE (ALICE, LAURA AND IRIS'S STORIES)

Located in a small Staffordshire village near Stone, a peaceful rural hamlet with a tiny population of about 900, the Swynnerton filling factory site was chosen as a munitions location because it was situated in an area prone to mist, making it difficult to observe from above. Construction of Swynnerton started in 1939.

By the summer of the following year, the factory was operational: shells and bombs which had been made in other munitions factories were being sent to Swynnerton for assembly.

Swynnerton's workforce grew rapidly: it had 5,000 workers in 1940, many of whom had been working at the Woolwich Arsenal in London and were relocated to the new facility.

By in the summer of 1941, the workforce totalled 15,000, reaching a peak of 18,000 a year later, the majority of whom were female. Very soon, the women working there were nicknamed The Swynnerton Roses.

Seven miles in radius, the Swynnerton site consisted of over 2,000 small buildings, separated from each other by considerable distances. The site had its own roadway system. For safety reasons, shell or detonator filling sites were surrounded by high banks of large mounds of earth and blast walls. These were specially constructed in order to reduce the risk of one explosion triggering off yet another explosion.

Decontamination centres were also constructed around the factory site in case of a German gas attack.

Mustard gas poisoning– which had been used by the Germans in WW1– was lethal. It could cause severe burning and blistering of the skin, blindness, external and internal bleeding. Victims often died a painful and lingering death. (Workers at Swynnerton recall special awareness sessions where demonstrations were given of a worker being painted all over with whitewash, believed to protect badly burnt skin).

The groups of different workshops or factories at Swynnerton were linked by a cleanway system. This consisted of raised walkways covered with a layer of smooth asphalt. These walkways were always kept scrupulously clean, to ensure that no grit or dirt was carried into the factory buildings.

The walkways had also been constructed with camouflage

in mind: from the air, the smooth asphalt surface looked just like water. So any German bombers flying over the site would not be able to see that below them lay a vast secret munitions complex. Yet despite such precautions, there were unsuccessful attempts to bomb Swynnerton and bombs did fall, on one occasion, on Swynnerton's railway lines. But thankfully, none were dropped on the buildings.

Swynnerton's roadway system was complemented by its own internal railway station. This station was constructed just outside the main factory gate at Cold Meece, a hamlet on the east side of the factory. It had four platforms. The double track at the Cold Meece line was used to transport workers to and from the area. 19 passenger trains a day ran on the line, Monday to Saturday. Yet the station never appeared on any public timetable and the factory never appeared on any Ordnance Survey maps until the Sixties. Cold Meece also used special trains to transport groups of American Airforce servicemen who were temporarily accommodated in the Swynnerton factory hostels.

Swynnerton's workforce consisted mainly of women ages 18-35. Although many workers were local girls, especially from the Potteries area fifteen miles away, there were large numbers of women from all over the country working there, many sent to live in Swynnerton's specially constructed hostel complex called Frobisher Hall. This consisted of seven prefabricated hostel buildings, all set up in the vicinity of the factory.

Filling detonators at Swynnerton was a highly dangerous task: so much concentration was required for the work, Swynnerton Roses were forbidden from even talking to each other as they worked.

Other workers recall specific tasks in the detonator section that were so dangerous, just one worker would be allocated to work on this kind of task – alone, in a locked room. In that way, if anything did go wrong, only one worker's life was placed at risk. You can only imagine the thoughts of these women at the time, though the super-risky task was only carried out once by those designated to do so.

One equally dangerous section at the Swynnerton site was called 1 West. It was nicknamed 'Suicide Section' because there were frequent accidents there. Many injured Swynnerton women were sent to the local North Staffs Hospital, where one ward was specially reserved for accidents from Swynnerton. Some women survived such accidents with terrible, extensive injuries: one pregnant woman lost both hands and her sight in an explosion, though the baby was born safely afterwards.

If an explosion took place and there were no injuries to workers, the women in the section were given an hour's break, two asprin from the site nurse and sent back to work. Working time was a precious commodity.

Although production of ammunition ceased at Swynnerton after war ended, the site remained as a factory until 1958. Many buildings had to be destroyed because of the toxic chemicals that had been used in them. As for the women who had worked there, they stepped back into normal life, raised their families. Many continued to work locally in the pottery industry. Today, the Swynnerton site is now an Army training base.

Only in 1991, forty-five years after the war ended, did the Swynnerton Roses' brave and valuable contribution to the war effort start to be openly recognized. This was kickstarted with a play, performed locally at Newcastle under Lyme. The play, called 'I don't want to set the world on fire' was written by Bob Eaton and was based on interviews and wartime memories of surviving Swynnerton Roses.

Perhaps the words from one of the songs in 'World on Fire' best sums up the Bomb Girls' experience at Swynnerton – and at all the other munitions factories too.

The Roses of Swynnerton
Work hard night and day
If it wasn't for the Roses
Where would old England be?'

(Quote from *I don't want to set the world on fire* by Bob Eaton, 1991)

AYCLIFFE, CO. DURHAM
(LAURA'S STORY)

In May 1940, construction work started on a large Royal Ordnance filling factory at Aycliffe in County Durham, near Bishop Auckland. The site was considered to be especially suitable because, like the other big filling factories, the area tended to be misty – which meant it was not easily seen from the air.

The factory started production in April 1941. Covering 867 acres, the Aycliffe factory consisted of over 1,000 buildings. At its peak production in 1943, Aycliffe employed 17,000 people working on shifts round the clock. 85 per cent of these were women, employed to put powder into shells and bullets or help assemble detonators and fuses. Many of these women had never worked in a factory before.

Despite every effort being made to conceal Aycliffe, the Germans were aware of its existence.

A pro-Nazi broadcaster, William Joyce, who had fled Britain in 1939 in order to escape internment, adopted the name 'Lord Haw Haw' and made it clear in his regular English language propaganda broadcasts from Germany to Britain and the U.S. that the Germans knew about the Aycliffe site – and the thousands of women who were working there.

Again and again, he broadcast chilling stories that the Luftwaffe would be bombing the factory out of existence, along with the Aycliffe girls.

'Those little angels of Aycliffe won't get away with it,' he claimed.

Yet it was an empty boast. On a few occasions, Aycliffe workers were strafed by enemy aircraft yet Aycliffe itself was never bombed. And the name 'Aycliffe Angels' stuck and was later adapted to 'White Angels' because the women wore white overalls as they worked. (After the war, Joyce was tried for treason against Britain. He was found guilty and met the hangman's noose).

Inside the Aycliffe factory there were twelve separate sectors, each one responsible for producing a different type of munition. During the wartime years, it is estimated that Aycliffe produced nearly one billion pieces of munition – including 700,000,000 bullets. The bullet production line had women filling bullet casings with cordite powder at one end and at the other end, the bullets were finished with machines placing the 'caps' on the bullets. Then the bullets were packed into boxes ready to be shipped off to the Armed Forces.

Many of the buildings at Aycliffe were small and well spaced out, to minimise the effect of any explosions. The buildings were of a special construction with a very strong framework and walls made of light infill brickwork, designed to just 'blow out' in the event of an explosion, leaving the main structure of the building intact.

The roadways in the site (known as 'cleanways')were wide, to enable staff to move materials around the site as safely as possible. Those running north to south were known as Avenues, those running east to west were called Streets.

Over half the women working at Aycliffe were married. Many had young children. However, there was no nursery or crèche at Aycliffe, though some of the other Royal Ordnance factories did provide these facilities for mothers.

Typically, Aycliffe operated round the clock in a three-shift pattern, with some shorter shifts available for women with young families.

Aycliffe munitions workers were not housed in specially built hostels, though some were billeted in nearby Darlington.

Many of the workers travelled in by bus or train each day from a radius of about 25 to 30 miles. Some 'secret housing' to accommodate factory workers and their families was built by the Ministry of Supply in 1942. It comprised some 350 houses in the Eastbourne area of Darlington – but very little was known about it at the time. Plans to build more housing were drawn up – but were shelved.

The average age of the women was 34, but over 1,000 of the Aycliffe Angels were over 50. One of these women, Mary Alice Dillon from Crook, was awarded the British Empire Medal for services to the country.

Mary had claimed to be 49. In fact, she was 69, with an excellent attendance record. In her two and a half years spent working at the factory, Mary missed just two shifts.

Like Bridgend, Aycliffe was split into two halves. There was the 'Clean Side' where volatile explosive compounds were found in loose open quantities in pots, on tables. Safety regulations here were draconian.

The other, 'Dirty Side', where the dangerous compounds were not in evidence in the same way, was equally safety conscious – but the safety regulations were slightly more relaxed. Special clothing had to be worn to work on the 'Clean Side' and this had to be removed before crossing back to the 'Dirty Side'.

Factory workers were repeatedly reminded of the three key safety factors they had to adhere to: strong discipline, absolute routine and precision in their work. Safety slogans abounded on the factory floor. One memorable slogan said:

'A concealed mistake is a crime. It may cost not only your life – but the lives of others.'

It is not known exactly how many lives were lost whilst Aycliffe was in production during the war years, though there were some recorded major explosions and fatalities. Accidents in some areas were common and despite all the regulations, some were caused by carelessness. One young girl was killed because she left a hair clip in her hair. It fell into the machine she was working on and caused an explosion.

One of the worst accidents at Aycliffe happened in February 1942 when 200lbs of fulminate exploded. The explosion killed four factory workers. The force of the explosion was so great, their bodies were virtually shredded of flesh.

Eight more Aycliffe workers lost their lives and one injured in another explosion in May 1945, just days before the war ended. The explosion was so loud, it could be heard for some miles around the factory.

The conflict between production targets and safety regulations caused much tension. The need to produce the munitions was relentless and each section or team had very challenging targets to meet. Bonuses were paid when production targets were met – so there was a temptation sometimes to cut corners. Supervisors needed to be aware of workers putting pressure on other workers to 'get more material through', so that the team qualified for their bonus.

The downside, of course, was that cutting corners in such a dangerous environment could prove fatal. For instance, a rush to meet the tight targets could sometimes cause staff to take short cuts by not cleaning machines thoroughly. This could result in too much loose powder being present in a machine. And this would cause an explosion.

One Aycliffe Angel worked with a man who had been in the Navy. He had been torpedoed. One day, the man spotted one of his co- workers finishing a bomb with a faulty fuse. He went berserk and had to be removed from the area. He knew all too well the danger of one faulty bomb during wartime.

Workers had to volunteer to work in the most dangerous area of the factory, known as Group 1. Like Swynnerton, this danger zone was nicknamed 'the suicide group'. This was the initiator group where the highly sensitive explosives were mixed together to create a more solid material. All accidents were shocking but serious accidents in the 'mixing shop' were truly horrific.

As dangerous as it was to manufacture the bombs or mix the explosives, it was equally hazardous to move the raw ingredients and components of the bombs around the site.

Most of the content of a shell or bomb is a high explosive chemical. In itself, the chemical is unlikely to explode without the presence of another chemical – the detonator. The detonator must be powerful and capable of exploding easily if it is given a blow or shock of some kind.

As a consequence, working with detonators and carrying them around the site was potentially lethal. Workers were not, under any circumstances, permitted to carry detonators through doorways –in case the door swung back and hit the load. So the detonators had to be lifted through hatchways in the side of the buildings.

Taking the detonators to be tested meant an employee with a red flag would walk 30 yards in front of the two people carrying the detonators. Workmen on the road who saw this thought it was a huge joke. What they didn't know was that the detonators were capable of blowing everyone and everything sky high.

Sorting the chemicals before they were used created another big safety issue.

For instance, fulminate of mercury is safe if it is kept damp. Yet to use it in the production of shells and bombs, it needs to be dry. So the chemical had to be dried on tables, in special buildings called 'drying houses'. It was carried around the site in small paper cups.

Another main chemical, lead azide, was more dangerous when it was wet than when it was dry. So moving this chemical around on a wet day was extremely hazardous.

ROF Aycliffe ceased to operate as a munitions site in August 1945. After the war the site was converted into an industrial estate renamed Newton Aycliffe and the area itself became a new town in 1947.

As hard, dangerous and draining as the work was, after the war the women who had worked there remembered their lighter moments the lunchtime concerts, the canteen lunches and, most importantly, the friendships and close bonds they formed at the factory. Many went on to last a lifetime.

Today, Newton Aycliffe is the second largest industrial estate in the north east. Some of the blast walls and buildings that housed the workers can still be seen, proud testament to the work of the courageous Aycliffe Angels.

DRUNGANS, DUMFRIES, SCOTLAND (MARGARET PROUDLOCK'S STORY)

The Drungans site at Cargenbridge, Dumfries, is the largest industrial complex in the Dumfries and Galloway region.

In 1939, it was selected by ICI and the Government as a suitable location for the manufacture of acids and nitro-cellulose for explosives.

Land that had previously been used for farming was to be transformed into a top secret industrial site. The area was relatively safe from bombing, yet within reach of ICI's factory and laboratory at Ardeer in Ayrshire.

The factory went into production in January 1941, a considerable achievement since all the construction work had to be carried out in daylight (the blackout regulations meant that lights could not be used on the building at night).

Apart from the construction of more than 40 buildings for production, storage, offices, fire station and canteen, a mile of roads and 1,900 yards of railway line were also laid.

Workers at the site came from as far away as Dundee (150 miles away). Many were billeted with local people in the area. At peak production, the site had a workforce of 1,350: more than half of the workers were women.

Through the war, more than 1.1 million tons of acid were produced at the Drungans factory. 37,500 tons of guncotton were delivered to nearby factories at Powfoot and Dalbeattie to be turned into cordite.

The factory closed down briefly in 1945 but it was reopened by ICI the following year to produce sulphuric acid for industry.

It was eventually rebuilt as a purpose built factory, initially manufacturing synthetic fibres, then it continued as a manufacturing plant for a wide range of industrial materials ranging from high quality film to packaging material. In 1998 ICI's 60 year link to Drungans ended when the site was taken over by the international DuPont Teijin corporation.

LEVER BROTHERS/PORT SUNLIGHT, THE WIRRAL, LIVERPOOL.
(IVY GARDINER'S STORY)

The vast Port Sunlight factory and housing village in the Wirral Peninsula was created in 1887 by the Lever Brothers, William Hesketh Lever and James Darcy Lever as a model village. At first, the company made branded soap and detergent products, then it expanded into the manufacture of margarine and ice cream products until the Twenties when it merged with a Dutch company to form Unilever, the first ever modern multinational.

The major feature of the company, right from the start, was that the Port Sunlight village adjoining the factory was built to accommodate the staff in good quality housing with extensive community and leisure facilities.

Employees and employers alike were regarded as one large family, with a workforce growing up together in a tradition of 'enlightened industrial outlook.'

As a consequence of this, working at Port Sunlight was regarded as a bright option around the Liverpool area, when Ivy Gardiner first joined the company as a 15-year-old filling soap packets.

In August 1940, direct war work started at Port Sunlight with the manufacture of small parts and the reconditioning of blitzed machine tools from firms in the area.

Though not a purpose built munitions complex, Port Sunlight was a bombing target because of its proximity to the port of Liverpool, so strategically important for Britain.

In October 1940, the Sunlight village was severely damaged during an air raid where three people were killed. Soap production was severely reduced with the introduction of soap rationing in 1942, releasing more space and manpower for war work. By then, US Army Jeeps and giant Dodge trucks were arriving by sea in parts, in huge cases, ready to be assembled at Port Sunlight and eventually used in the North Africa campaign.

Port Sunlight's biggest wartime engineering undertaking, The Dowty Department, was also set up in 1942 to manufacture retractable under carriages, designed by George Dowty, for Lancaster Bombers. Working on the assembly of these under- carriages was the war work Ivy Gardiner trained for at age 21. In February 1943, the first of these under carriages left the factory to be assembled with the Lancaster bombers, ready for combat.

After the war, the factory resumed normal production. All soap production there ceased in 2001 but today, the model village at Port Sunlight is a heritage area and museum. The adjoining Unilever factory is scheduled to be extended to a high-tech factory site for personal care products like deodorant and shampoo.

ROF BLACKPOLE, WORCESTER
(MAISIE JAGGER'S STORY)

Located a mile to the north of Worcester Shrub Hill Station on the Worcester to Birmingham railway line, the Blackpole site was originally a Government owned munitions factory during the First World War. It was known as Cartridge Factory No. 3 and run by Kings Norton Metal.

In 1921, the site was purchased by Cadbury Brothers, Bournville as additional factory space until 1940 when it was requisitioned by the Government as a small arms ammunition factory producing cartridge cases.

During wartime the site had its own railway station, Blackpole Halt, to transport workers to and from the factory. The site was handed back to Cadburys in 1946.

For many years, Cadburys Cakes were produced at Blackpole until Cadbury's merger with Schweppes in the Seventies when the site was sold. It is now a retail park.

BISHOPTON, RENFREWSHIRE
(MARGARET CURTIS' STORY)

Bishopton was chosen as a munitions site for the manufacture of explosives because of its favourable micro-climate. It was also in Clydeside, an area of high unemployment in the Thirties, and had good rail links.

Part of the site was constructed on requisitioned farm land; the southern end of the site had originally been a filling factory during World War 1, employing over 10,000 workers.

Three self-contained explosive manufacturing factories were built on the Bishopton 2,000 acre site. Construction started in 1937 and the facility, built specifically to manufacture propellant, mainly cordite, for the Army and RAF, opened at the end of 1940. The site had its own bus service and internal railway lines used to transport explosives around the site. At its peak, Bishopton employed 20,000 workers, most of them female.

Each building on the site was numbered, with Factory 0 housing the supporting services for the site. (These included a permanently manned fire station with its own fire brigade, ambulance station, medical centre, mortuary, laboratories, clothing and general stores, machine shops, general workshops and laundry).

The three main factory buildings (1, 2 and 3) each had their own coal fired power stations. Factories 1 and 2 had their own nitration plant for making nitrocellulose (known as gun cotton) which was then processed on site to produce cordite. Each factory at Bishopton had its own nitroglycerine section.

Factory 3 closed down almost immediately after WW2. The manufacture of explosives including cordite, gunpowder and other explosives continued at Bishopton in Factories 1 and 2 until the year 2000. The site is now owned by BAE Systems.

Ordnance Survey maps did not show the existence of Bishopton until after the year 2000.

Long after WW2, it remained, as did so many other wartime munitions sites, top secret.